PATRICK MOORE'S
A-Z of Astronomy

PATRICK MOORE'S
A-Z of Astronomy

Patrick Stephens, Wellingborough

First published in 1986

*British Library Cataloguing in Publication
Data*

Moore, Patrick
　Patrick Moore's A-Z of astronomy.
　1. Astronomy—Dictionaries
　I. Title
　520'.3'21　　QB14

　　ISBN 0-85059-800-1

*Patrick Stephens Limited is part of the
Thorsons Publishing Group*

Photoset in 9 pt Times by MJL Typesetting
Services Limited, Hitchin, Herts. Printed in
Great Britain on 115 gsm Fineblade coated
cartridge, and bound, by Anchor Brendon
Limited, Tiptree, Colchester, Essex, for the
publishers, Patrick Stephens Limited,
Denington Estate, Wellingborough,
Northants, NN8 2QD, England.

Preface

The last edition of this book appeared some years ago. When I came to revise it, I realized that it needed to be completely rewritten in view of the rapid developments in astronomy which have taken place recently, and this I have done, at the same time taking the opportunity to extend the text and make it more comprehensive.

I am most grateful to Paul Doherty for his usual excellent illustrations, to John Mason and Barney D'Abbs for their helpful comments and to Darryl Reach and Bruce Quarrie, of Patrick Stephens Limited, for all their help and encouragement.

Patrick Moore
Selsey, July 1985

A

Aberration of starlight The apparent displacement of a star from its true position in the sky, due to the fact that light has a definite velocity (186,000 miles per second) and does not move at infinite speed. A good analogy is to picture a man walking along in a rainstorm, holding up an umbrella to shield himself. To keep dry, he will have to slant the umbrella

Aberration.

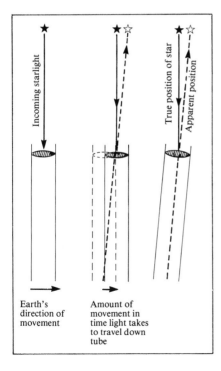

Incoming starlight

True position of star

Apparent position

Earth's direction of movement

Amount of movement in time light takes to travel down tube

forward, as shown in the diagram; in other words, the raindrops will seem to come in at an angle instead of straight down. In the case of starlight, the aberration effect is due to the movement of the Earth, which is travelling round the Sun at an average velocity of 18½ miles per second; thus the starlight seems to arrive at an angle. The true positions of stars may be affected by up to 20.5 *seconds of arc. The aberration of starlight was discovered by James *Bradley; it was the first actual demonstration of the Earth's movement round the Sun.

Absolute magnitude The *magnitude that a star would seem to have if it were observed from a standard distance of 10 *parsecs, or 32.6 *light-years.

Absorption of light in space It was formerly thought that space must be completely empty. This is now known to be wrong; there is a vast amount of thinly-spread interstellar matter, so that the light coming from remote objects is partially absorbed. For example, we cannot see the centre of the *Galaxy at all, though we can study it by infra-red and radio waves which are not blocked by the interstellar dust.

Acamar The star Theta Eridani. As its *declination is 40°S, it is not visible from Britain. The magnitude is 2.92 and the distance 55 light-years. It has been suspected that Acamar has faded from the first magnitude over the past 2,000 years, but the evidence is far from conclusive. It is a fine double; the components are rather unequal, and the separation is 8.5 *seconds of arc.

Accretion disk A disk of material from which larger bodies may be formed under the influence of gravitation. Also the rotating disk of material formed around a *Black Hole, detected by the copious X-rays which are emitted.

Achernar The star Alpha Eridani; for data see *Stars. Achernar is never visible from Britain or the United States, and is the nearest really bright star to the *south celestial pole; its *declination is 57°S.

Achilles One of the *Trojan asteroids. Its magnitude is 15.3, and its period 11.9 years.

Achondrite A type of stony *meteorite, containing very little iron or nickel.

Achromatic lens A lens corrected for *chromatic aberration, so that 'false colour' is reduced. An achromatic lens is made up of two components made of different types of glass, whose errors tend to cancel each other out.

Adams, John Couch (1819-1892) English mathematical astronomer, chiefly remembered for his correct prediction of the planet *Neptune—though the actual discovery of Neptune was made from similar calculations by the French astronomer *Le Verrier.

Adams, Walter S. (1876-1936) American astronomer; Director of the *Mount

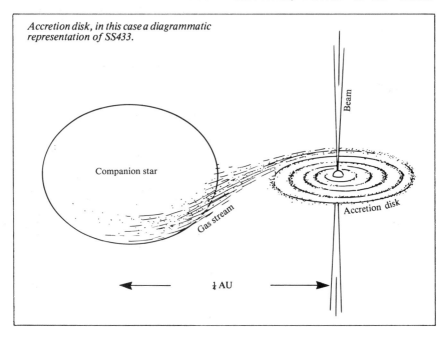

Accretion disk, in this case a diagrammatic representation of SS433.

Companion star

Gas stream

Beam

Accretion disk

¼ AU

Wilson Observatory from 1923 to 1946. His main work was in stellar spectroscopy.

Aerolite A stony *meteorite.

Airglow The faint luminosity of the night sky, due mainly to processes going on in the Earth's upper atmosphere.

Airy disk The apparent size of a star's disk produced by a perfect optical system. Since the star can never be focused perfectly, 84 per cent of the light will concentrate into a single disk, and 16 per cent into a system of surrounding rings.

Airy, Sir George Biddell (1801-1892) The seventh Astronomer Royal, who was appointed to the post in 1835 and retired in 1881. He was a great administrator, though decidedly autocratic. He made many contributions to astronomy and timekeeping, and was responsible for raising *Greenwich Observatory to a position of eminence.

Aitken, Robert (1864-1949) A great American observer of double stars, and one-time Director of the *Lick Observatory.

Albedo The reflecting power of a planet or other non-luminous body; the ratio of the amount of light reflected from the body, to the amount of light which falls upon the body from an outside source. A perfect reflector would have an albedo of 100 per cent. The average albedo of the surface of the Moon is only 7 per cent!

Albireo The star Beta Cygni. It is a lovely double, with a yellow primary and a vivid blue companion; the magnitudes are 3.2 and 5.4 respectively, and as the separation is 34.6 *seconds of arc the pair may be split with a very small telescope.

Alcor The star 80 Ursæ Majoris; the naked-eye companion of *Mizar.

Alcyone The brightest star in the *Pleiades cluster. It is of magnitude 2.87, and is 350 times as luminous as the Sun; spectral type B.

Aldebaran The star Alpha Tauri; for data see *Stars. Aldebaran is orange, and has been nicknamed 'the Eye of the Bull'. It appears to lie in the star-cluster of the *Hyades, but it is not a true member; it simply happens to lie about half-way between the Hyades and ourselves.

Algol The prototype *eclipsing binary. The magnitude range is from 2.1 to 3.3, with a period of 2.69 days; the secondary minimum is very slight. Eclipses of the primary by the larger, dimmer secondary are not total. The descent to minimum takes 4 hours; minimum lasts for 20 minutes, and the recovery to maximum takes a further 4 hours. The variability of Algol was discovered by Montanari in 1669. Though Algol is called 'the Demon Star', and lies in the head of the Gorgon, Medusa, it is not now believed that the ancients knew about its unusual behaviour.

Alioth Epsilon Ursæ Majoris, one of the seven stars of the *Plough. For data see *Stars.

Alkaid Eta Ursæ Majoris, one of the seven stars of the *Plough but not a member of the *moving cluster. For data, see *Stars.

Allende Meteorite A *carbonaceous chondrite which fell in Mexico in 1969, scattering 5 tons of material.

Almagest The great book by *Ptolemy of Alexandria, summarizing much of the scientific knowledge of the ancient world. It has come down to us by way of its Arab translation.

Alpha Centauri The nearest of the bright stars; for data, see *Stars. It is a fine binary; the components are of magnitudes 0.0 and 1.4, with a revolution period of 80 years and a separation ranging from 2 to 22 *seconds of arc. It is too far south to be seen from anywhere in Europe. It and Agena or Beta Centauri (magnitude 0.61) are the Pointers to the *Southern Cross.

Alphard The star Alpha Hydræ; for data,

see *Stars. Alphard is often called 'the Solitary One', because of its isolated position in the sky; its orange colour makes it easy to recognize. Sir John *Herschel believed it to be variable, but this has not been confirmed.

Alps, Lunar A range of mountains bordering the Mare *Imbrium, and cut through by the famous Alpine Valley.

Al-Sûfi (903-986) Arab astronomer, who compiled an important star-catalogue.

ALSEP Apollo Lunar Surface Experimental Package. Each successful *Apollo mission left an Alsep on the Moon, and data continued to be sent back until the equipment was deliberately switched off.

Altair The star Alpha Aquilæ. For data see *Stars. Altair is easy to recognize because it is flanked to either side by a fainter star (Gamma and Beta Aquilæ).

Altazimuth mount A type of telescope mount upon which the instrument may be swung freely in any direction (see dia-

Altazimuth mount showing directions of movement.

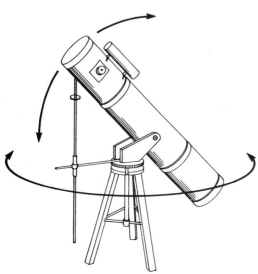

gram). Until recently large telescopes were not set up upon altazimuth mountings, owing to the difficulty of guiding; but modern computers have overcome this problem, and some major telescopes, such as the Russian 236-in, are on altazimuth mounts.

Altitude The angular distance of a celestial body above the horizon, ranging from 90° at the *zenith or overhead point down to 0° at the horizon.

Amalthea The fifth satellite of Jupiter, discovered by E. E. Barnard in 1892. For data see *Satellites. It was surveyed by the *Voyager space-craft, and found to have a reddish, cratered surface. It is decidedly irregular in shape. Amalthea was the last planetary satellite to be discovered visually.

Amirani An active volcano upon *Io.

Amor asteroids *Minor planets whose orbits bring them inside that of Mars, but without crossing that of the Earth. Amor itself was discovered in 1932; it has a period of 2.7 years, but is only a few miles in diameter. Its asteroid number is 1221.

Ananke The 12th satellite of Jupiter. (See *Satellites.)

Andromeda Galaxy The nearest of the really large external *galaxies. It is dimly visible with the naked eye; photographs show it to be spiral, though as it lies at an unfavourable angle to us the full beauty is lost. It lies at a distance of 2.2 million light-years, and is considerably larger than our own *Galaxy. It is known officially as M.31 (NGC 224). There are two important companions to it, M.32 and NGC 205, both of which are very easy telescopic objects.

Andromedid meteor shower The shower associated with the defunct *Biela's Comet; sometimes called the Bieliid shower. It reaches its maximum in late November, but has now become very feeble.

Ångström unit One hundred-millionth

part of a centimetre. The wavelength of visible light ranges between about 3900 Å (violet) and 7500 Å (red). The unit is named after the last-century Swedish physicist Anders Ångström.

Antares Alpha Scorpii; the 'Rival of Mars', so named because of its strong red colour. Its diameter is over 250,000,000 miles. For data see *Stars.

Antoniadi, Eugenios (1870-1944) Greek-born astronomer who spent much of his life in France, and became the most distinguished planetary observer of his time. He was particularly noted for his studies of Mars and Mercury, carried out with the 33-in refractor at the Observatory of *Meudon.

Apennines The most conspicuous mountain range on the Moon, bordering the Mare *Imbrium. Its peaks rise to

Right *The NGC 891 spiral nebula in Andromeda seen edge-on* (Mount Wilson and Palomar).

Below *A Lunar Rover in the foreground with Apollo 15 behind and the Apennines in the background.*

11

almost 15,000 ft. The *Apollo 15 lunar module landed in the foothills of the Apennines.

Aperture synthesis A technique used in *radio astronomy. Several 'dishes' are used, some of which are movable. By varying the positions of the 'dishes' and making use of the Earth's rotation, it is possible to obtain a resolving power equal to that of a much larger single instrument.

Aphelion The position in the *orbit of a planet or other body when furthest from the Sun. For instance, the Earth is at aphelion in early July, when its distance from the Sun is 94½ million miles. Similarly, *apogee refers to the furthest distance of a body moving round the Earth.

Aphrodite Terra The largest highland region on *Venus, near the planet's equator. It measures 6,000 miles by 2,000 miles, and consists of eastern and western mountains separated by a lower region. It adjoins *Atla Regio.

Apogee This is described under the heading *Aphelion.

Apollo asteroids *Minor planets whose paths cross that of the Earth. All are very small. No 1862, Apollo, was discovered in 1932, and then lost until picked up once more in 1973; it can make close approaches to the Earth, Venus and Mars. Many Apollo asteroids are now known. It has been suggested that they may be ex-comets which have lost all their volatiles, but the evidence is very uncertain. Other notable Apollo asteroids are Adonis, Toro, Geographos, Dædalus, Cerberus, Sisyphus and Midas.

Apollo programme The American manned lunar flight programme. The first landing was made by Neil *Armstrong

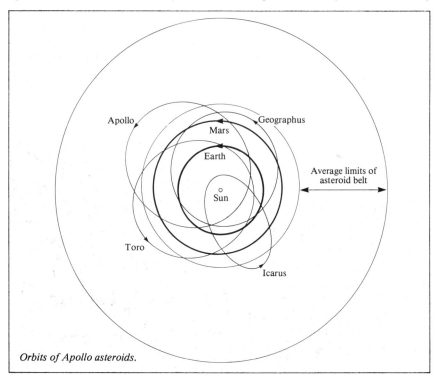

Orbits of Apollo asteroids.

The ascent stage of Apollo 11 carrying Neil Armstrong and Edwin Aldrin back to the command module following the first Moon landing.

and Edwin Aldrin (July 1969) in Apollo 11; subsequent landings were made by the crews of Apollos 12, 14, 15, 16 and 17. The programme ended with Apollo 17 in December 1972. Lunar samples were brought back for analysis, and *ALSEPS were left on the Moon's surface.

Appulse The apparent close approach of one celestial body to another. It does not indicate an actual close approach, but is a line of sight effect.

Arago, François (1786-1853) French astronomer, who became Director of the Paris Observatory in 1830. He was concerned largely with studies of the Sun, and also wrote popular books.

Arecibo radio telescope The world's largest radio telescope 'dish' built in a natural hollow in the ground at Arecibo, Puerto Rico. It is 1,000 ft in diameter. Obviously it is not steerable, but adjustments can be made by moving the antenna placed above the 'dish'.

Comet Arend-Roland, 1957.

Arend-Roland comet A bright comet seen

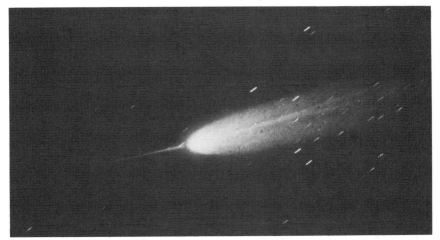

in 1957. It was at its best in April, and was a prominent naked-eye object, with a tail and a 'reverse tail' which was actually due to material, scattered in the comet's orbit, catching the sunlight at a favourable angle.

Areography The physical study of *Mars (from the Greek, *Ares*).

Argelander, Friedrich Wilhelm August (1799-1875) German astronomer who produced important star catalogues, notably the Bonner Dürchmusterung.

Argyre Planitia A well-marked basin in the southern hemisphere of *Mars.

Ariane The three-stage launch vehicle operated by the European Space Agency. In July 1985 the *Giotto probe to *Halley's Comet was launched by an Ariane rocket, from Kourou in French Guiana. Many commercial satellites have been launched by Arianes.

Ariel The inner large satellite of *Uranus.

Aristarchus, photographed by Apollo 15.

For data, see *Satellites. It is believed to be made up of a mixture of rock and ice.

Aristarchus The brightest crater on the Moon, 23 miles in diameter, with terraced walls and a central peak. It has often been mistaken for an erupting volcano, and when on the dark hemisphere is usually easy to see by *earthshine. Many *TLP or Transient Lunar Phenomena have been seen in and near it.

Aristarchus of Samos (*c* BC 310-250) Greek astronomer, who was one of the first, if not the very first, to maintain that the Earth is in orbit round the Sun. He also tried to measure the relative distances of the Moon and Sun, and his method was quite sound in theory, though inaccurate in practice.

Arizona Crater See *Meteor Crater.

Armagh Observatory The leading observatory in Northern Ireland, founded in 1790. It also has a major *planetarium.

Armstrong, Neil (1930-) The first man on the Moon; he landed in the Mare *Tranquillitatis in July 1969 from *Eagle*, the

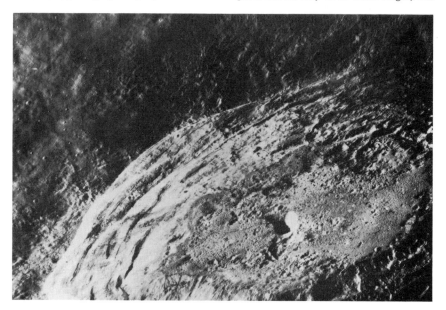

lunar module of Apollo 11. He had previously flown in the Gemini programme, and had joined the space programme in 1962. From 1971 to 1980 he was Professor of Engineering at Cincinnati University.

Arrhenius, Svante (1859-1927) Swedish chemist and Nobel Prize winner. He was the author of books on astrophysics, and was also the originator of the *panspermia theory.

Artificial satellites Man-made vehicles, put into *orbits round the Earth. An artificial satellite is launched by rocket; once it has been put into a stable orbit it will not come down unless it spends any part of its time within the resisting part of the atmosphere, in which case it will be affected by friction against the air-particles and will have its orbit gradually changed until it re-enters the lower air and is destroyed. Of course, an artificial satellite may be brought down to a controlled landing, as has been the case with many unmanned vehicles—and, needless to say, with all manned ones!

The first artificial satellite, Sputnik 1, was launched by the Russians on 4 October 1957; this may be said to mark the real beginning of the Space Age. Many more have followed, and by now many nations have joined in the programmes.

Artificial satellites have been put to many uses. They have been able to provide invaluable information about Earth resources; they are used as communications relays, and for all sorts of scientific investigations. Meteorological satellites have led to a great improvement in weather forecasting, and there have also been cases in which satellites have sent back early warnings of developing storms, so that thousands of lives have been saved. Unfortunately there are also many satellites used for purely military purposes—something which all true scientists will deplore.

Some of the satellites have been brilliant—notably the two *Echo balloon vehicles of the 1960s. Most, however, look only like slowly-moving stars.

Ascending node This is described under the heading *Node.

Asgard Large ringed formation on *Callisto.

Ashen Light When the planet *Venus appears as a crescent, the 'night' side sometimes appears dimly luminous. This is known as the Ashen Light. It is almost certainly a genuine phenomenon, probably due to electrical phenomena in the upper atmosphere of Venus. There is no comparison with the well-known *earth-shine on the Moon, since Venus has no satellite.

Asteroids See *Minor planets.

Astræa Minor planet No 5, discovered in 1845. It is considerably smaller and fainter than the first four known members of the swarm. For data see *Minor planets.

Astrograph A telescope designed specifically for astronomical photography.

Astrolabe An instrument used by ancient astronomers to measure the *altitudes of celestial bodies. An astrolabe consists of a circular disk marked off in *degrees along its rim; the object is sighted by means of a movable arm, with the astrolabe held suspended vertically, and the altitude is then read off upon the scale.

Astrology The pseudo-science which attempts to link human characters and destinies with the positions of the planets against the starry background. It has no scientific foundation whatsoever, and has long since been completely discredited.

Astrometric binary A *binary system in which the fainter component cannot be seen, but makes its presence felt by its gravitational effects upon the proper motion of the visible star.

Astrometry The branch of astronomy dealing with the movements and positions of celestial bodies.

Astronautics The science of space research, using either unmanned or manned vehicles. It is a modern development, since it is only during the last few decades that rockets have become powerful scientific tools, but it has now become of fundamental importance, and has provided information which could not possibly have been obtained in any other way.

Astronomical twilight The period between sunset and the time when the Sun has dropped to 18° below the horizon.

Astronomical unit The distance between the Sun and the Earth; in round figures 93,000,000 miles. The mean distance from the Earth to the sun is now known to be slightly less than this (92,957,209 miles or 149,598,500 km), but for most purposes it is good enough to regard the astronomical unit as 93,000,000 miles.

Astronomy The science dealing with the bodies in the sky. It began in prehistoric times, when our remote ancestors gazed up at the stars and divided them into *constellations, but its real development began with the Greeks, who drew up excellent star catalogues, measured the size of the Earth, and studied the movements of the Sun, Moon and planets. However, most of the Greeks made the fundamental mistake of assuming the Earth to lie in the centre of the universe.

*Ptolemy, last of the great astronomers of ancient times, died about AD 180, and for some centuries after this little progress was made. A revival came with the Arabs of a thousand years ago, and eventually astronomy returned to Europe. In 1546 *Copernicus published a book in which he claimed that the Earth is merely a planet moving round the Sun. Arguments about this went on for more than a century, but the work of men such as *Galileo and *Kepler, followed by *Newton's researches into the nature of gravitation, finally provided the answer.

Telescopes were invented in the early 17th century, and were used astronomically by Galileo and others as early as 1610. Though he was not the first telescopic astronomer, Galileo was certainly the most skilful, and he made a series of spectacular discoveries—such as the satellites of Jupiter, the phases of Venus, spots on the Sun and the countless stars of the Milky Way.

With the construction of more powerful telescopes, progress became rapid, and in the 19th century came the development of spectroscopy and photography. From about 1870 large *refractors were built, and photography became all-important, eventually replacing the human eye for most branches of research. During the 20th century, large *reflectors superseded the refractors. *Radio astronomy began in the 1930s, though it was not until after the end of the war that it became of fundamental importance. The Space Age began in 1957, with the launching of the first *artificial satellite, and space research methods have led to the development of what is known as 'invisible astronomy'—*infra-red, *X-rays, *gamma-rays and so on. Most of the radiations coming from space are blocked by the Earth's atmosphere, so that instruments carried aloft in satellites and space-probes have to be used. We are also experiencing the 'electronic revolution', in which electronic devices are taking the place of photography.

Yet it is also true that even today, astronomy is still a science in which amateurs can play a valuable rôle, and there is close collaboration between amateur and professional astronomers.

Astrophysics The branch of modern astronomy which deals with the physics and chemistry of the stars. See also *Stars and *Spectroscope.

Ataxite A type of iron *meteorite, rich in nickel.

Aten asteroids *Minor planets whose orbits lie largely inside that of the Earth. Aten itself, discovered in 1976, was the first known asteroid to have a revolution period of less than a year (0.95 years); its distance from the Sun ranges between 0.79 and 1.14 *astronomical units. Aten asteroids are less common than those of the *Amor or *Apollo type. The first three to be found were Aten, *Ra-Shalom and *Hathor.

Atla Regio One of the main actively volcanic regions on Venus.

500 km	
Exosphere	
400 km	
Thermosphere	
200 km	
100 km	
Mesosphere	
50 km	
Stratosphere Ozone — 25 km	
10 km	
Troposphere	

Ionosphere

Left *Cross-section of the atmosphere.*

Atmosphere The gaseous mantle surrounding a planet or other body. The Earth's atmosphere is made up of several layers; the bottom seven miles or so comprises the *troposphere, above which come more rarefied layers such as the *stratosphere, *ionosphere and finally the *exosphere, which has no definite boundary, but simply 'tails off' into space. Though the atmosphere extends upward for hundreds of miles, most of its mass is concentrated in the bottom four or five miles.

The Earth, with an *escape velocity of 7 miles per second, has been able to retain a dense atmosphere. Bodies with lower escape velocities, such as Mars (3.1 miles per second) have thinner atmospheres, while the Moon (escape velocity 1½ miles per second) has none. On the other hand, the giant planet Jupiter (escape velocity 37 miles per second) has been able to hold down even hydrogen, the lightest of all the gases.

It is assumed that the Earth's original atmosphere, made up largely of hydrogen, escaped in the early history of the Solar System, so that our present atmosphere was produced by gases sent out from below the crust by volcanic action.

Atom The smallest unit of a chemical *element which retains its own particular character. Each atom may be said to consist of a central nucleus around which are more particles known as *electrons; the nucleus has a positive electrical charge and the electrons have negative charges, the two balancing each other and making the whole atom electrically neutral. Modern atomic theory is highly complicated, and it is quite misleading to suppose that an atom is a miniature Solar System, made up of solid lumps, but it is almost impossible to give an accurate description in non-mathematical language. There are 92 known naturally-occurring atoms, and it is safe to say that none has been overlooked. Extra atom-types have been made artificially, but all are heavier than the last 'natural atom', uranium.

Atoms link up to form atom-groups or

17

*molecules; thus a molecule of water is made up of two hydrogen atoms combined with one oxygen atom, so that its chemical formula is the familiar H_2O.

Aurora Auroræ are the lovely Northern Lights (Aurora Borealis) and Southern Lights (Aurora Australis). They occur in the Earth's upper atmosphere, and are due to charged particles sent out by the Sun, which penetrate the outer air and produce the beautiful glows. Auroræ are linked with events taking place on the Sun; a brilliant solar *flare, which emits charged particles, is quite likely to be followed by a major auroral display about 24 hours later.
   Because the particles are electrified, they tend to move toward the Earth's magnetic poles. Consequently, auroræ are best seen in high latitudes. In North Norway, Iceland and even North Scotland they are common during the hours of darkness; in South England, brilliant displays are rare, and in low latitudes auroræ are almost (though not quite) unknown. Obviously, auroræ are commonest when the Sun is at its most active.

Australite A *tektite from the Australian tektite field.

Azimuth The angular bearing of an object in the sky, measured from north (0°) through east (90°), south (180°) and west (270°) back to north (360° or 0°). Due to the Earth's rotation, the azimuths and *altitudes of all celestial bodies change constantly.

B

Baade, Walter (1893-1960) A German-born astronomer who spent much of his life in the United States. He made many notable contributions to stellar astronomy, and discovered the error in the *Cepheid scale which led to a doubling of the estimated distances of the *galaxies.

Babcock, Harold Delos (1882-1968) American astronomer, who spent most of his career at *Mount Wilson Observatory. His major contributions were in spectroscopy and in studies of the Sun, particularly solar magnetism. In his later life he collaborated closely with his son, H. W. Babcock.

Baily, Francis (1774-1844) English amateur astronomer, best remembered for his observations of *Baily's Beads at the total solar eclipses of 1836 and 1842.

Baily's Beads Brilliant points seen along the edge of the Moon's dark disk at a total *solar eclipse, just before and just after actual totality. They are caused by the Sun's light shining through valleys on the limb of the Moon, between mountainous regions.

Ball, Sir Robert Stawell (1840-1913) Irish astronomer, who worked for a time with Lord *Rosse at Birr Castle. In 1874 he was appointed Astronomer Royal for Ireland. He made many contributions to stellar and planetary astronomy, but is perhaps best remembered for his popular books, such as *The Story of the Heavens*, which were regarded as standard works.

Balloon astronomy The technique of sending up instruments in balloons, to heights of 85,000 ft or so, above most of the Earth's atmosphere. It has been of great importance, though by now it has been largely superseded by space research methods.

Barnard, Edward Emerson (1857-1923) American astronomer, noted for his comet discoveries, his planetary observations, and his work upon dark nebulæ. In 1892 he discovered *Amalthea, the fifth satellite of Jupiter.

Barnard's Star Munich 15040, a red dwarf star in the constellation of Ophiuchus. It is only 6 light-years away, and is therefore the nearest star beyond the Sun apart from the *Alpha Centauri system. Studies by P. van de Kamp and his colleagues at the Sproule Observatory, in America, indicate that Barnard's Star may be attended by an orbiting planet or planets. The star has the largest proper motion known—10.3 *seconds of arc per year, so that in 180 years it covers a distance equal to the diameter of the full moon; but it is 'weaving' its way along, and planetary companions may be responsible. Barnard's Star itself has only 1/2,000 the luminosity of the Sun.

Barwell Meteorite A meteorite which landed near Barwell, in Leicestershire, on 24 December 1965; during its fall it was widely observed, and many fragments were recovered. It was a stony meteorite, or *aerolite.

Barycentre The centre of gravity of the Earth-Moon system. The Earth is 81 times more massive than the Moon, so that the barycentre lies inside the Earth's globe.

Bayer, Johann (1572-1625) A Bavarian lawyer and amateur astronomer. In 1603 he published a star atlas, in which he introduced the system of allotting Greek letters to the stars in each constellation—a system which is still followed.

Becklin-Neugebauer object (Often termed BL.) An infra-red source deep in the *Orion Nebula, M.42. It cannot be seen optically, and is generally thought to be a very massive and luminous star, though its precise nature is still a matter for debate.

Beer, Wilhelm (1797-1850) German banker and amateur astronomer, who colla-

Fragments of the Barwell Meteorite, with a coin for comparison.

borated with J. H. von *Mädler in compiling the first really good map of the Moon.

Bellatrix The star Gamma Orionis. For data, see *Stars.

Bessel, Friedrich (1784-1846) German astronomer, who made many contributions but is best remembered as being the first man to measure the distance of a star (61 Cygni, in 1838).

Beta Canis Majoris variables See *Beta Cephei variables.

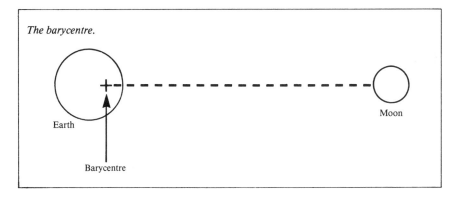

The barycentre.

Earth

Moon

Barycentre

Drawing of Biela's Comet in 1846 (A. Secchi).

Beta Cephei variables (Also known as Beta Canis Majoris variables.) Pulsating, very hot stars with very short periods and very small magnitude ranges. Beta Cephei and Beta Canis Majoris are the best-known members of the class.

Beta Regio A volcanic highland on Venus, containing two large shield volcanoes, Rhea Mons and Theia Mons, which are almost certainly active.

Betelgeux Alpha Orionis. For data, see *Stars. Betelgeux is a huge red supergiant, with a diameter of over 250,000,000 miles; it is variable between about magnitudes 0.2 and 0.8. Using the technique of *speckle interferometry, it has been claimed that patches have been recorded on its surface, though this interpretation has been challenged. Though lettered Alpha, Betelgeux is less brilliant than *Rigel, or Beta Orionis.

Biela's Comet A comet which used to move round the Sun in a period of 6¾ years. At the return of 1845 it split in two; the 'twins' were seen for the last time in 1852. In 1872 and 1885 a brilliant meteor shower was seen instead of the comet, and a few of these *Andromedid meteors are still seen every November. There is no reasonable doubt that Biela's Comet is defunct.

Big Bang theory The theory that the universe came into existence at one set moment in time, between 15,000 and 20,000 million years ago.

Binary star A star made up of two com-

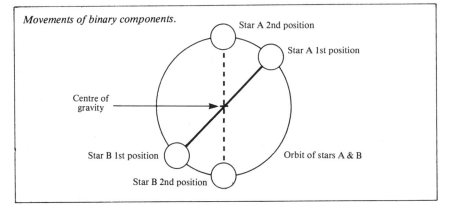

Movements of binary components.

Star A 2nd position

Star A 1st position

Centre of gravity

Star B 1st position

Orbit of stars A & B

Star B 2nd position

ponents, genuinely associated, and moving round their common centre of gravity. Binaries are very common in space. In some cases the components are equal; such is Theta Serpentis, which is easily split with a small telescope. With other pairs, such as *Mizar in the Plough, the components are unequal. There are also many binaries in which one component far outshines the other; such is *Sirius, which is accompanied by a *White Dwarf companion with only 1/10,000 the luminosity of its primary.

The periods of revolution have considerable range. With pairs which are widely separated, the period may amount to millions of years, and all we can really say is that the components share a common motion in space; very close pairs have short periods. When the separation is very small indeed, the components cannot be detected individually; pairs of this sort are termed *spectroscopic binaries. With some systems the components are virtually touching each other.

There are also many 'stellar families', or multiple stars, such as *Castor in Gemini, which is made up of two bright stars, each of which is again double, plus a dim double red dwarf.

It used to be thought that a binary star must be the result of the break-up or fission of a formerly single star, but it is now believed that the components merely formed from the same nebulosity in the same region of space.

Binoculars A pair of binoculars consists of two small *refractors joined together, so that the observer can use both eyes. The aperture in millimetres and the magnification determine the type of binoculars: thus 7 × 50 indicates a magnification of 7, with each *object glass 50 mm in diameter. In many ways binoculars are much more useful astronomically than very small telescopes such as 2½ in refractors and 4 in reflectors, though with magnifications of above about ×12 some sort of mounting is desirable. Comet-hunters often make use of specially-mounted, large-aperture, wide-field binoculars.

BL See *Becklin-Neugebauer object.

Black body A body which absorbs all the radiation which it receives; that is to say, it has an *albedo of zero. It is also a perfect emitter of radiation. It is, of course, a theoretical concept only.

Black Drop An appearance seen at the start of a *transit of Venus. As the planet moves on to the Sun's disk it seems to draw a strip of blackness after it; when this strip disappears, the transit has already begun—making it impossible to time the moment of ingress accurately. The effect is caused by the dense atmosphere surrounding Venus.

Black Dwarf A dead star, which has exhausted all its reserves of energy. Whether the universe is yet old enough for any Black Dwarfs to have formed is by no means certain.

Black Hole A localized region of space from which not even light can escape, because of the presence of an old, collapsed star or *collapsar.

If a star is too massive to end its career as a *White Dwarf or as a *neutron star, it may form a Black Hole. The *escape velocity becomes greater than the velocity of light, and the star will be unobservable; it will be surrounded by a 'forbidden zone' which is to all intents and purposes cut off from the rest of the universe. The boundary of this zone is known as the *event horizon*. Inside the event horizon, all the ordinary laws of science break down; it has even been suggested that the old star will crush itself out of existence altogether.

Obviously we can track down Black Holes only by their gravitational effects upon bodies which we can see—in fact, if they are members of *binary systems. Perhaps the most plausible case is that of Cygnus X-1, which is made up of a supergiant star, HDE 226868, together with a massive component which is probably a Black Hole. X-rays are being received, presumably due to material which is about to pass over the event horizon and is being very strongly heated. This material has formed an *accretion disk around the Black Hole.

It must be stressed that there is no

positive proof that Black Holes exist, though most astronomers are confident that they do. Some bizarre suggestions have been made—such as the idea that a Black Hole might be an entry point to another part of the universe or even to a completely different universe. This sounds remarkably like science fiction. Yet the whole concept of Black Holes is extraordinary by any standards, and we can only await the results of future research.

Blink-comparator See *Blink microscope.

Blink-microscope (or Blink-comparator) An instrument for examining two photographs in rapid succession. If the same star-field is photographed at different times, and the pictures are shown in rapid succession—three or four 'blinks' per second—a moving object, such as an asteroid, will seem to jump to and fro, thereby betraying its nature. Various bodies, such as *Pluto, have been identified in this way. Blink-microscopes are also used for studies of the *proper motions of stars.

Bliss, Nathaniel (1700-1764) The fourth Astronomer Royal. He held office only

A Bok Globule.

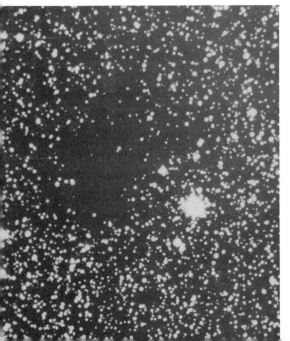

from 1762 to his death two years later.

BL Lacertæ objects Very active galaxies with condensed centres, which in some ways resemble *quasars, though they are much less powerful and have different spectra. BL Lacertæ itself was identified in 1968; previously it had been taken to be an ordinary variable star. It is widely believed that the main energy source of a BL Lacertæ object is a central *Black Hole.

Bode's Law An interesting relationship between the distances of the planets from the Sun, first noticed by J. D. Titius in 1772 and popularized by J. E. Bode. It may be summed up as follows:

Take the numbers 0, 3, 6, 12, 24, 48, 96, 192 and 384, each of which (apart from the first) is double its predecessor. Add 4 to each. Taking the Earth's distance from the Sun as 10, the distances of the other planets known in Bode's time are given with reasonable accuracy, as follows:

Planet	Distance Bode's Law	Actual
Mercury	4	3.9
Venus	7	7.2
Earth	10	10
Mars	16	15.2
Jupiter	52	52.0
Saturn	100	95.4

Uranus, discovered in 1781, was found to have a distance of 191.8; Bode's Law had predicted 196, while the gap corresponding to the missing number 28 was filled by the *minor planets, of which the first, *Ceres, was discovered in 1801 and fitted well into the scheme. However, Neptune, discovered in 1846, departs from the rule, as its distance is only 300.7 instead of the predicted 388; the last number corresponds more closely to Pluto (distance 394.6), which has however an unusually eccentric orbit and may not be worthy of planetary rank.

On the whole, it seems that Bode's Law is merely coincidence, and has no real significance.

Bok globule A small dark object seen against a background of stars or a gaseous *nebula. The globules are named in

Years

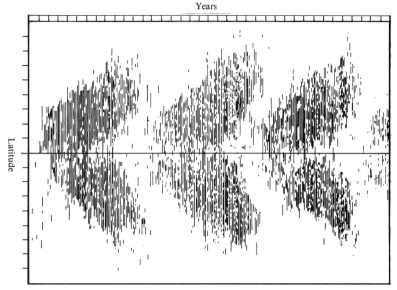

Latitude

Butterfly diagram.

honour of the Dutch astronomer Bart J. Bok, who first drew attention to them. They may well be protostars, not yet hot enough to shine.

Bolide A brilliant *meteor, which may explode during its descent through the Earth's atmosphere.

Bolometer A very sensitive electrical instrument, used to measure tiny quantities of heat radiation.

Boyden Observatory A major South African observatory, at Bloemfontein. The largest telescope is a 60-in reflector.

Bradley, James (1693-1762) The third Astronomer Royal, from 1742 to his death. He compiled a very accurate star catalogue, and also discovered the *aberration of starlight.

Brahe, Tycho See *Tycho Brahe.

Breccia A rock formed by the sudden fusing of various fragments of rock under pressure. Many lunar breccias have been obtained from the samples brought back by the Apollo astronauts and the Russian unmanned probes.

Brorsen's Comet A short-period comet with a period of 5.5 years. It was seen at five returns, the last being in 1879, but it has now disappeared, and is almost certainly defunct.

Brown, Ernest William (1866-1938) English mathematician with a particular interest in celestial mechanics; he produced the standard tables of the Moon's motion.

Burnham, Sherburne Wesley (1838-1921) American astronomer, who worked at the *Lick and *Yerkes Observatories. He specialized in double star research, and discovered over 1,300 new pairs, as well as producing a standard catalogue of double stars.

Butterfly diagram A graphical representation of *Spörer's Law, which states that the new spots of a *sunspot cycle appear in high latitudes and that as the cycle progresses, the spots appear in latitudes closer and closer to the Sun's equator. The graph gives the impression of successive pairs of butterfly wings!

C

Calendar The system of dividing time into convenient periods—such as days, weeks and months—for our everyday needs. The obvious basic unit is the time taken for the Earth to go once round the Sun, generally termed the 'year', but this raises difficulties, because the true 'year' is not an exact number of days. While completing a full journey round the Sun, the Earth spins on its axis 365¼ times, but to have a calendar taking quarter days into account would be absurd. Therefore, we take the year to be 365 days, and add on an extra day every fourth year to make up for the difference; this every fourth year is a Leap Year, with the shortest month, February, having 29 days instead of its usual 28.

An easy way to tell which years are Leap Years is to divide by 4. If there is no remainder, then the year is a Leap Year. The only exceptions are the 'century' years (1700, 1800, 1900 etc) in which the division is by 400. Thus 1900 was not a Leap Year, but 2000 will be.

Various calendars have been used in the past. Our own is adapted from that of the Romans. Originally there were only ten months, of which the first was March; Julius Cæsar introduced the 'Julian calendar', which was first used in BC 44 and was a notable improvement; a further modification was made in 1582 by order of Pope Gregory XIII. Britain adopted the Gregorian Calendar in 1752.

Callisto The fourth satellite of Jupiter. For data, see *Satellites.

Callisto is almost as large as the planet Mercury, but is much less massive, and is the faintest of the four main satellites of

Callisto, a photo-mosaic made up from Voyager photographs in 1979.

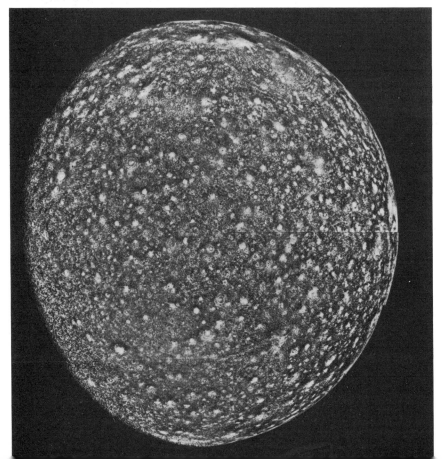

Jupiter, though any small telescope will show it. It is believed to have a silicate core, overlaid by a mantle of water or soft ice which is in turn overlaid by the icy crust. Callisto was surveyed in detail by the *Voyager space-craft, and we have accurate maps of most of its surface. Callisto is more heavily cratered than any other world known to us; there are also two large ringed basins, *Valhalla (diameter 370 miles) and the similar though smaller Asgard. There is no evidence of past tectonic activity, so that Callisto must have been inert since the early days of the Solar System. Like all major planetary satellites, Callisto has a *captured or synchronous rotation: 16.7 Earth-days, so that the same face is always turned toward Jupiter.

Caloris Basin (More properly, Caloris Planitia.) A formation on Mercury, photographed from Mariner 10 in 1974. It is 800 miles in diameter, and is bounded by a ring of mountain blocks. Unfortunately only part of it was in sunlight during the three active passes of Mariner 10. It lies near a 'hot pole', the highest temperature region of Mercury—hence the name.

Caloris Planitia See *Caloris Basin.

Campbell, W. W. (1862-1938) American astronomer; Director of the *Lick Observatory, 1901-1931. He was a pioneer in studies of the *radial velocities of stars, but also made major contributions to planetary research, particularly in connection with Mars.

Canals, Martian Straight, artificial-looking features on Mars recorded by astronomers such as G. V. *Schiaparelli and Percival *Lowell. They were first drawn in detail by Schiaparelli in 1877; Lowell was of the opinion that they were artificial waterways. However, the space-craft pictures show that the Martian canals do not exist in any form; they were purely illusory.

Canaveral, Cape The American rocket base in Florida. At one stage it was renamed Cape Kennedy, but the change was unpopular, and before long the original name was restored. It was from here that most of the major US space missions have been launched.

Cannon, Annie Jump (1863-1941) American woman astronomer, who worked at the Harvard College Observatory and was concerned mainly with stellar spectra—in which her contributions were outstanding.

Canopus Alpha Carinæ; apart from *Sirius, the brightest star in the sky. It is very luminous and remote; unfortunately it is too far south to be seen from Britain. For data, see *Stars.

Cape Observatory The oldest South African observatory. It is still the main administrative centre, though the main telescopes have now been moved to a more favourable site at *Sutherland.

Capella Alpha Aurigæ, almost overhead as seen from Britain during winter evenings. For data, see *Stars.

Captured rotation (or synchronous rotation) If the axial rotation period of a body is equal to its revolution period, the rotation is said to be captured. Tidal forces over the ages are responsible. The Moon has captured rotation, so that it always keeps the same face turned Earthward; the same is true of all other large planetary satellites.

Carbon-nitrogen cycle The stars are not 'burning' in the usual sense of the term; they are producing their energy by nuclear reactions. Normal stars contain a great deal of hydrogen, and the hydrogen nuclei are being converted to helium nuclei, with release of energy and loss of mass. One way in which this conversion takes place is by a whole series of reactions, in which two more elements, carbon and nitrogen, are concerned; this is the carbon-nitrogen cycle. It was once thought that the Sun shone by this process, but it is now known that another cycle, the *proton-proton cycle, is more important in stars of solar type. However, the end result is the same: hydrogen is converted into helium, and the star continues to shine.

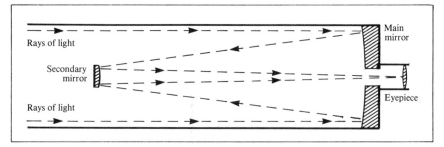

Principle of the Cassegrain reflector.

Carbon stars Red stars of spectral types R and N, containing an unusual amount of carbon in their atmospheres.

Carme The eleventh satellite of Jupiter. For data, see *Satellites. Carme moves in a retrograde direction, and may well be a captured asteroid.

Carrington, Richard (1826-1875) Pioneer English amateur observer of the Sun; he observed the first solar *flare in integrated light.

Cassegrain reflector A type of reflecting telescope in which the main mirror has a central hole. The light from the object to be observed passes down the telescope tube and is reflected from the main mirror on to a smaller, convex mirror termed the secondary. The light is then sent back through the hole, and the image produced is magnified by an *eyepiece in the usual way. With a Cassegrain, therefore, the observer looks up the instrument instead of into the tube, as with a *Newtonian reflector.

Cassegrains have many advantages, and are becoming more and more popular; but although the tube is shorter than that of an equal-aperture Newtonian, the telescope is less easy to make.

Cassini division The principal division in Saturn's ring system, separating Rings A and B. It is visible with a small telescope when the rings are suitably placed. It is not empty; the *Voyager probes have shown that it contains narrow ringlets, not visible from Earth.

Cassini, Giovanni (1625-1712) (Often known as Jean Dominique Cassini; though Italian, he spent much of his life in France.) First Director of the Paris Observatory. He was a great observer of the planets, and discovered four of Saturn's satellites as well as the main division in the ring-system.

Cassiopeia A A strong radio source in the constellation of Cassiopeia. It is certainly a *supernova remnant, and it is thought that the outburst may have dated from about 1702, though it was unobserved because of intervening obscuring material. It is over 9,000 light-years away, and is also a source of X-rays.

Castor Alpha Geminorum; for data, see *Stars. Castor is a fine binary, with a period of 350 years. Each component is a *spectroscopic binary, and there is also a third member of the system, YY Geminorum, which also is a spectroscopic binary—so that the Castor system consists of six suns. Early astronomers made it brighter than its 'twin', *Pollux; it is now fainter, though it is hardly likely that there has been any real alteration.

Catadioptic telescope A type of telescope which uses both refraction and reflection to form the image at the prime focus. The errors in the primary mirror are remedied by a special correcting plate covering the full aperture.

CCD *Charge coupled device.

'Celestial police' A group of astronomers, who met in 1800 to hunt for a new planet between the orbits of Mars and

Jupiter. The prime movers were J. H. *Schröter and the Baron Franz Xavier Von *Zach. The first asteroid, *Ceres, was discovered in 1801 by G. *Piazzi, who was not then a member of the team, but in 1802 Ceres was recovered by H. *Olbers, one of the 'police', and three more discoveries followed: *Pallas, *Juno and *Vesta. No more asteroids seemed to be forthcoming, and the 'police' disbanded.

Celestial sphere An imaginary sphere surrounding the Earth. For this purpose we may suppose that the sky is solid (as the ancients used to believe), and that we lie in the exact centre of the sky-sphere. A useful, though imperfect, way to show what is meant is to picture a table-tennis ball suspended in the exact centre of a football; the table-tennis ball then represents the Earth, while the inside surface of the football represents the celestial sphere.

The Earth spins on its axis once a day (roughly 24 hours). The axis points northward to the *north celestial pole, marked approximately by the bright star *Polaris in Ursa Minor; there is no bright star at the *south celestial pole. As the Earth rotates, the celestial poles seem to remain stationary, while the rest of the sky moves round. From northern latitudes, the south celestial pole can never be seen,

The celestial sphere.

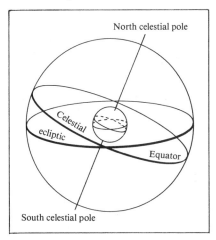

North celestial pole

Celestial
ecliptic

Equator

South celestial pole

as it always stays below the horizon; similarly, observers south of the equator can never see Polaris. The celestial sphere is divided into two equal hemispheres by the *celestial equator.

Centaurus A The nearest major radio galaxy: NGC 4151. It is 13,000,000 light-years away, and is highly complex. It was once believed to be made up of two galaxies in collision, but this is now known to be wrong. Centaurus A is also a strong X-ray source.

Cepheid An important type of *variable star. Cepheids have short periods of from a few days to a few weeks; they are perfectly regular, and it has been found that the real luminosity of a Cepheid is linked with its period of variation. The longer the period (that is to say, the interval between one maximum and the next), the more luminous the star; thus a Cepheid with a period of 7 days will be more powerful than a Cepheid whose period is only 6 days. From this, it follows that once a Cepheid has been studied, and its period measured, its real luminosity can be found, which in turn allows its distance to be calculated. Of course, many corrections have to be made, notably to allow for the *absorption of light in space, but the basic principle is clear enough.

Cepheids are brilliant, pulsating stars, far more luminous than the Sun, so that they are visible across great distances. In 1923 E. E. *Hubble detected Cepheids in the *Andromeda Galaxy, and found them to be so remote that there could no longer be any doubt that the Andromeda Galaxy itself lay far beyond our Milky Way system.

It has since been shown (by W. *Baade) that there are two distinct classes of Cepheids, with different period-luminosity relationships, but all can act as effective 'standard candles'. Bright naked-eye members of the class are Delta Cephei (the prototype star, after which the class has been named); Eta Aquilæ, Zeta Geminorum, and the southern Beta Doradûs.

Ceres The largest and first-discovered of the *minor planets. Its distance corres-

Sizes of Ceres and Britain/Ireland compared.

ponds well enough with the 'missing' *Bode figure 28. Ceres never becomes bright enough to be seen with the naked eye; it is of the carbonaceous type.

Cerro Tololo Observatory A major observatory near La Serena in Chile, equipped with 158-in and 60-in reflectors. It is run by a group of American universities.

Challis, James (1803-1862) Director of the Cambridge Observatory, unfortunately best remembered for his failure to discover the planet *Neptune even though all the information had been put into his hands.

Chandrasekhar limit The limiting mass for a *White Dwarf star; it is equal to 1.44 times the mass of the Sun. If the mass is greater than this, the collapsing star will become a *neutron star or even a *Black Hole, as the material will not be able to support itself against the tremendous gravitational force. The name honours the Indian astrophysicist S. Chandrasekhar.

Chao Meng Fu A crater on *Mercury, containing the Mercurian south pole.

Charge Coupled Device (Often termed a CCD.) A very sensitive electronic device, far more effective than a photographic plate. Used with a large telescope, a CCD can record objects far fainter than any previously detected.

Charon The satellite of *Pluto, described under that heading.

Chiron A strange object discovered in 1977 by Charles Kowal with the Schmidt telescope at *Palomar. It has been given an asteroid number (2,060) but is unique inasmuch as it moves mainly between the orbits of Saturn and Uranus. Its period is 50.7 years, and the distance from the Sun ranges between 8.5 and 18.5 *astronomical units. The next perihelion will be in 1996, when the magnitude will rise to about 15. Chiron is large by asteroid standards, and one estimate gives the diameter as about 400 miles, though this is very uncertain. Its nature is still uncertain, but it seems to have a low *albedo, in which case it is not an icy globe like most of the satellites of Saturn. It has been traced on photographs taken as long ago as 1895.

Christie, Sir William (1845-1922) The

Chiron, 19 October 1977, photographed through the 48-in Schmidt telescope at Palomar.

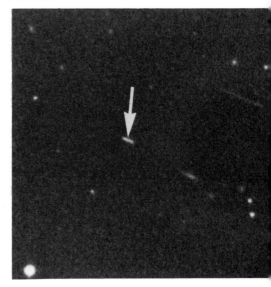

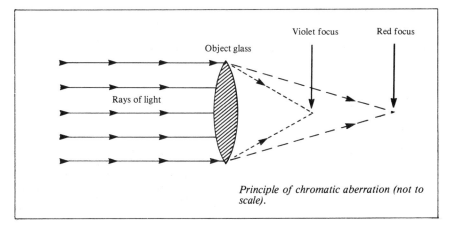

Principle of chromatic aberration (not to scale).

eighth Astronomer Royal; he succeeded *Airy in 1881, and retained the post until retiring in 1910.

Chromatic aberration A defect found in all lenses, resulting in the production of 'false colour'. Light is a mixture of various wavelengths; for visible light, violet has the shortest wavelength (3,900 *Ångströms) and red the longest (7,500 Ångströms). When passed through a lens, the shorter wavelengths will be the more strongly bent or *refracted*, so that violet will be brought to focus closest to the lens. In fact, the various component colours are brought to focus in different places, so producing a whole series of images instead of only one; the result is an annoying amount of spurious colour.

By using a compound lens, made up of several separate lenses of different kinds of glass, chromatic aberration may be reduced, but it can never be entirely cured. In a refracting telescope, the

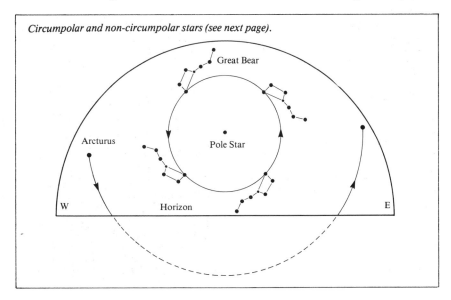

Circumpolar and non-circumpolar stars (see next page).

The Pleiades and associated nebulosity in Taurus photographed with the 48-in Schmidt at Palomar.

*object-glass is always compound, and is known therefore as an *achromatic objective.

Chromosphere The part of the Sun's atmosphere lying above the bright surface or *photosphere, and below the *corona. It is visible with the naked eye only during a total solar *eclipse. By spectroscopic methods it may however be studied at any time. It is red, and is made up chiefly of hydrogen gas. The name really means 'coloursphere'.

Chronometer A very accurate form of timekeeper. The first chronometers were developed by the English clockmaker, Harrison, for use at sea.

Chryse Planitia The 'Golden Plain' on Mars; site of the first Viking landing on 16 July 1976. Chryse lies in the northern hemisphere of the planet, roughly between Aurorae Planum and Acidalia Planitia.

Circumpolar star A star which never sets, but merely circles the pole above the horizon. From Britain, the Great Bear is circumpolar, because it never dips out of view, whereas the brilliant orange star Arcturus is not (see diagram on previous page). From New Zealand, the Great Bear can never be seen, but the *Southern Cross is circumpolar.

Clark, Alvan (1808-1887) American optician who made many of the largest object-glasses, notably the *Yerkes 40-in.

Clouds of Magellan See *Magellanic Clouds.

Cluster-Cepheids Obsolete name for *RR Lyræ variables.

Clusters, Stellar As the name suggests, a cluster is a group of stars whose members are genuinely associated. They are of two main types: *open* and *globular*.

The open clusters are formless, and contain a few dozen or up to a few hundred stars. The most famous are the *Pleiades or Seven Sisters, in Taurus; the *Hyades, also in Taurus; *Præsepe in Cancer, and the lovely *Jewel Box, Kappa Crucis in the Southern Cross. The stars in an open cluster have a common origin. Eventually, gravitational forces will cause the cluster to disperse and lose its separate identity.

Globular clusters are regular in shape, and contain many more stars—over a million in some cases. They form a sort of 'outer framework' to the *Galaxy. The three brightest are *Omega Centauri and 47 Tucanæ in the southern hemisphere, and M.13 Herculis in the north. External galaxies also contain globular clusters.

Globular clusters belong to the galactic *halo, and are very ancient, so that their

M.67—old open cluster in Cancer.

leading stars have already evolved off the
*Main Sequence. Most of them contain
*Cepheid variables, by which their dis-
tances can be measured.

Coal Sack The famous dark *nebula in the
Southern Cross, too far south to be visible
from Britain or any part of Europe.

Cœlostat A form of optical instrument
which makes use of two mirrors, one fixed
and one movable. The movable mirror is
mounted parallel to the Earth's axis, and
is rotated so that the light from the object

under observation is reflected in a fixed
direction on to the second mirror. The
result of this arrangement is that the
*eyepiece does not have to move at all.

Strictly speaking, the term 'cœlostat'
applies only to this moving-mirror device,
but it has many applications. It is used, for
example, in *tower telescopes, designed
for studying the Sun, in which the
cœlostat is placed at the top of the tower
and the image of the Sun is formed at the
base, always in the same position.

Collapsar The end product of a very
massive star which has collapsed to form a
*Black Hole.

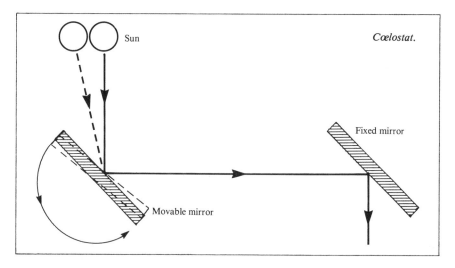

Colour index A measure of a star's colour, and hence of its surface temperature. The ordinary or *visual magnitude* of a star is a measure of its apparent brightness as seen with the eye; its *photographic magnitude* is obtained by measuring the size of the star's image on a photographic plate. The two magnitudes will not generally be the same, because different colours affect the sensitive plates in different ways; in most cases red stars will seem fainter photographically than they do to the eye. The difference between visual and photographic magnitude is the colour index. The scale is adjusted so that for a white star of spectral type A0 the colour index is zero. Blue stars have negative colour index; with red stars, the colour index is positive.

Colures Great circles on the *celestial sphere. The *equinoctial colure* is the great circle of *right ascension 0 hours and 12 hours; it passes through the celestial poles, the *First Point of Aries, and 180° of celestial longitude. The *solstitial colure* is the great circle of right ascension 6 hours and 18 hours, passing through the celestial poles, the poles of the *ecliptic, and the solstitial points (celestial longitude 90° and 270°, or right ascension 6 hours and 18 hours, and *declination 23½°N and 23½°S).

Coma There are two astronomical meanings here:
1. The hazy-looking patch surrounding the nucleus of a *comet.
2. The blurred haze surrounding the images of stars on a photographic plate, due to instrumental defects.

There is also a constellation, Coma Berenices (Berenice's Hair).

Comes ('Companion') The fainter component of a *binary system.

Comet A member of the Solar System. Most comets have very eccentric orbits, and apart from *Halley's Comet all the really bright comets seen over the ages have periods of many centuries—so that we never know when or where to expect them, and they are seen only at one return in many human lifetimes.

A comet is not a solid, rocky body. According to the accepted theory, due to

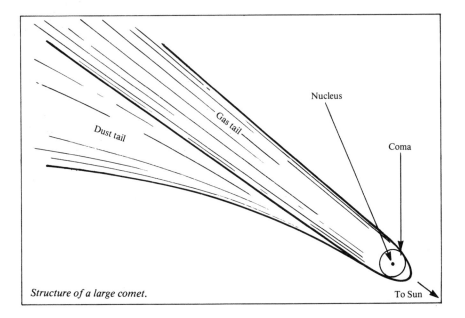

Structure of a large comet.

F. L. *Whipple, a comet may be described as a dirty ice-ball. The only reasonably massive part is the nucleus, made up of ices in which are embedded rocky particles. When the comet approaches the Sun, the ices in the nucleus start to boil off; a head or *coma is formed, hiding the nucleus completely. A large comet may also have a tail—or, rather, two tails, one made of gas (the 'ion tail') and the other of 'dust'. Generally the ion tail is straight, while the dust tail is curved. Comet tails always point more or less away from the Sun, due to the *solar wind (gas tails) and solar radiation pressure (dust tails).

However, not all comets develop tails, and many of the smaller comets look like nothing more than tiny patches of luminous haze in the sky. Comets depend upon sunlight to make them shine—mainly by reflection, though when near *perihelion the cometary material, particularly the ion tail, may be excited and therefore emit a certain amount of light on its own account.

There are many short-period comets, with their *aphelia near the orbit of Jupiter, but all these are faint, and very few can be visible with the naked eye even when at their best. The comet with the shortest known period is Encke's (3.3 years), which can in fact be followed all round its orbit. Comets of medium period are less common; the most famous is, of course, *Halley's (76 years).

Brilliant comets were seen quite frequently during the 19th century: for instance in 1811, 1843, 1858 (Donati's Comet), 1861, 1862, 1874 and 1882. Our own century has been poor in them; the last really splendid comet was that of 1910 (the so-called Daylight Comet, seen shortly before Halley's), though there are several which have attained naked-eye visibility; for example Arend-Roland (1957), Ikeya-Seki (1975), Bennett (1970) and West (1976). Kohoutek's Comet of 1973 was a great visual disappointment. It was expected to become brilliant, but signally failed to do so, though it was clearly visible with the naked eye; however, it was scientifically important as it was studied by the crew of the US space-station *Skylab and found to be surrounded by an immense hydrogen

Comet West, March 1976 (Richard West).

cloud. It will not return for about 75,000 years.

Comets are of very low mass, and are easily perturbed by the planets, particularly Jupiter. Since they lose some material at each return to perihelion, they must be relatively short-lived and we do indeed know of several periodical comets, seen during the last century, which have now disintegrated (*Biela's Comet being the most famous example). Comets are associated with *meteors, which may be described as cometary débris.

There used to be widespread alarm at the appearance of a bright comet, partly on superstitious grounds and partly because it was feared that a collision with a comet might destroy the Earth. This is, of course, out of the question, and even a

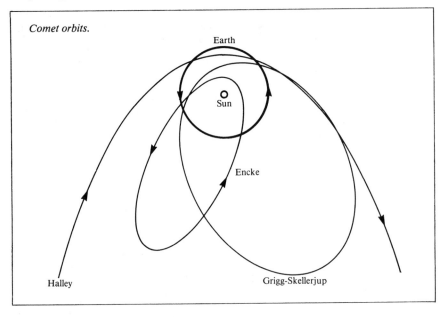

Comet orbits.

Earth

Sun

Encke

Halley

Grigg-Skellerjup

direct hit from a cometary nucleus would do no more than local damage, though it is true that the effects upon climatic conditions might be very marked (it has even been suggested that the extinction of the dinosaurs, about 65,000,000 years ago, was caused by the impact of a comet or an asteroid). In 1910 the Earth may have passed through the tail of Halley's Comet without any detectable effects.

Comets are believed to come from the *Oort Cloud, a swarm of comets orbiting the Sun at a distance of at least a light-year. If a comet is perturbed for any reason, it swings inward toward the Sun. If it is 'captured' by a planet, it may be forced into a short-period orbit; otherwise it will return to the Oort Cloud. If the orbit is perturbed in such a way as to turn it into an open curve (parabola or hyperbola), the comet will be expelled from the Solar System altogether.

Comets are usually named after their discoverers, though in some cases after the mathematician who first worked out the orbit. The following list gives some of the comets which have been seen at more than one return:

Comet	Period in years
Encke	3.3
Grigg-Skjellerup	5.1
Tempel 2	5.3
Tempel-Swift	5.7
D'Arrest	6.2
Pons-Winnecke	6.3
Kopff	6.4
Giacobini-Zinner	6.5
Borrelly	6.8
Finlay	6.9
Faye	7.4
Comas Solà	8.6
Wild	13.3
Tuttle	13.8
Crommelin	27.9
Tempel-Tuttle	32.9
Olbers	69.5
Pons-Brooks	71.0
Halley	76.1
Herschel-Rigollet	156.0
Grigg-Mellish	164.3

Compton Effect An interaction between a *photon and a charged particle such as an electron; there is a transfer of energy from the photon to the electron, so that the photon is re-radiated at a lower frequency. If photons of low frequency are scattered by moving electrons, they are re-radiated at higher frequency; this is the *Inverse Compton Effect.*

Conduction The transfer of heat by energy being passed directly from atom to atom.

Conjunction Here again there are several meanings of the term to be considered.

A planet is said to be in conjunction with a star when it passes close by; of course this is purely a line of sight effect. Planets may also be in conjunction with each other, or with the Moon.

Mercury and Venus, which are closer to the Sun than we are, are said to be at *inferior conjunction when they are more or less between the Sun and the Earth, and at *superior conjunction when they are on the far side of the Sun; the lining-up is not usually exact, because both Mercury and Venus have orbits which are appreciably inclined to that of the Earth. An exact lining-up at inferior conjunction results in a *transit.

The remaining planets lie outside the orbit of the Earth, and are said to be at superior conjunction when on the far side of the Sun; they are then above the horizon only during broad daylight, and are to all intents and purposes unobservable.

Constellation A group of stars named after a living or a mythological character, or an inanimate object; the names are generally used in their Latin forms, so that, for instance, the Great Bear becomes Ursa Major.

In ancient times there were various constellation systems; those of the Chinese, the Egyptians and so on. We use those which were drawn up by the Greeks. *Ptolemy, last of the great astronomers of the Greek school, listed 48 constellations, among which were the familiar Ursa Major, Orion and others. All Ptolemy's groups are still in use, but the list has been extended, because Ptolemy's maps could not cover the whole of the sky; he could know nothing about the far southern stars, which never rose above the horizon

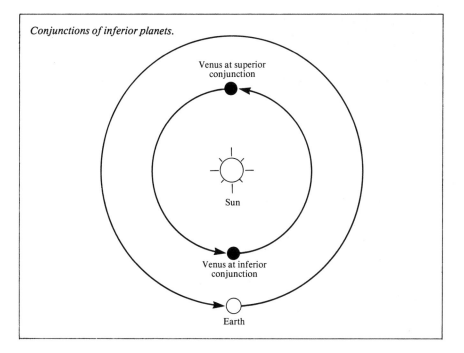

Conjunctions of inferior planets.

Venus at superior conjunction

Sun

Venus at inferior conjunction

Earth

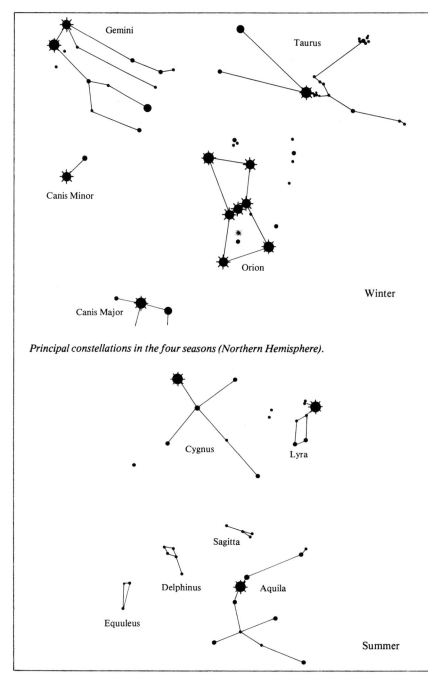

Principal constellations in the four seasons (Northern Hemisphere).

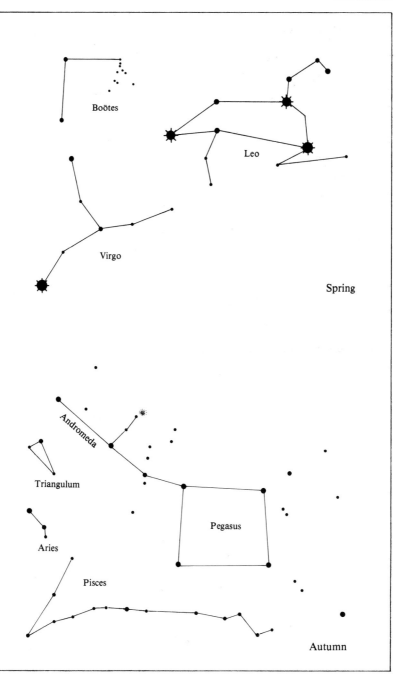

Boötes

Leo

Virgo

Spring

Andromeda

Triangulum

Aries

Pisces

Pegasus

Autumn

The constellation Orion.

of his home in Alexandria. Moreover, his original groups have been considerably modified.

A constellation is merely a line of sight effect; the stars in any particular constellation are not genuinely associated. Neither can it be said that many of the constellations have outlines which give a real impression of the objects they are supposed to represent. Ursa Major is nothing like a bear, and its nicknames of the Plough and (in America) the Big Dipper are more appropriate. Triangulum (the Triangle) is one of the few exceptions.

Because the stars are so remote, their individual or *proper motions are very slight, even though they are really moving through space at great speeds. Therefore, the constellation patterns seem to stay virtually unchanged over long periods.

The following constellations are now recognized by the *International Astronomical Union. Many others have been rejected, and the huge, unwieldy Argo Navis (the Ship Argo) has been chopped up into a keel, sails and a poop! Large and important constellations are given in capital letters.

Latin name	English name	1st-magnitude star(s)
ANDROMEDA	Andromeda	
Antlia	The Air-Pump	
Apus	The Bee	
AQUARIUS	The Water-bearer	
AQUILA	The Eagle	Altair
Ara	The Altar	
Aries	The Ram	
AURIGA	The Charioteer	Capella
BOÖTES	The Herdsman	Arcturus
Cælum	The Sculptor's Tools	
Camelopardalis	The Giraffe	
Cancer	The Crab	
Canes Venatici	The Hunting Dogs	
CANIS MAJOR	The Great Dog	Sirius
Canis Minor	The Little Dog	Procyon
Capricornus	The Sea-Goat	
CARINA	The Keel	Canopus
CASSIOPEIA	Cassiopeia	
CENTAURUS	The Centaur	Alpha Centauri, Agena
Cepheus	Cepheus	
Cetus	The Whale	
Chamæleon	The Chameleon	
Circinus	The Compasses	
Columba	The Dove	

Latin name	English name	1st-magnitude star(s)
Coma Berenices	Berenice's Hair	
Corona Australis	The Southern Crown	
Corona Borealis	The Northern Crown	
Corvus	The Crow	
Crater	The Cup	
CRUX AUSTRALIS	The Southern Cross	Acrux, Beta Crucis
CYGNUS	The Swan	Deneb
Delphinus	The Dolphin	
Dorado	The Swordfish	
Draco	The Dragon	
Equuleus	The Little Horse	
ERIDANUS	The River	Achernar
Fornax	The Furnace	
GEMINI	The Twins	Pollux
GRUS	The Crane	
Hercules	Hercules	
Horologium	The Clock	
Hydra	The Watersnake	
Hydrus	The Little Snake	
Indus	The Indian	
Lacerta	The Lizard	
LEO	The Lion	Regulus
Leo Minor	The Little Lion	
Lepus	The Hare	
Libra	The Balance	
LUPUS	The Wolf	
Lynx	The Lynx	
LYRA	The Lyre	Vega
Mensa	The Table	
Microscopium	The Microscope	
Monoceros	The Unicorn	
Musca Australis	The Southern Fly	
Norma	The Rule	
Octans	The Octant	
OPHIUCHUS	The Serpent-bearer	
ORION	The Hunter	Rigel, Betelgeux
Pavo	The Peacock	
PEGASUS	The Flying Horse	
PERSEUS	Perseus	
Phoenix	The Phœnix	
Pictor	The Painter	
Pisces	The Fishes	
Piscis Australis	The Southern Fish	Fomalhaut
PUPPIS	The Poop	
Reticulum	The Net	
Sagitta	The Arrow	
SAGITTARIUS	The Archer	
SCORPIUS	The Scorpion	Antares
Sculptor	The Sculptor	
Scutum	The Shield	
Serpens	The Serpent	
Sextans	The Sextant	

Latin name	English name	1st-magnitude star(s)
TAURUS	The Bull	Aldebaran
Telescopium	The Telescope	
Triangulum	The Triangle	
Triangulum Australe	The Southern Triangle	
Tucana	The Toucan	
URSA MAJOR	The Great Bear	
Ursa Minor	The Little Bear	
VELA	The Sails	
VIRGO	The Virgin	Spica
VOLANS	The Flying-Fish	
Vulpecula	The Fox	

Convection The transfer of energy by moving currents in gases or liquids.

Copernican sytem The system in which the Sun is the central body, with the Earth and other planets moving round it.

Copernicus, Nicolas (1473-1543) (The Latinized version of the real name, Mikolaj Kopernik.) Polish cleric and astronomer, who in 1543 published his great book *De Revolutionibus Orbium Cœlestium* (Concerning the Revolutions of the Celestial Bodies), in which he described the theory that the Sun is the centre of the Solar System. He withheld publication until the last days of his life, because he believed—correctly!—that the Church would be bitterly hostile.

Cor Caroli The star Alpha Canum Venaticorum, magnitude 2.9. It is a binary; the brighter component is the prototype *spectrum variable (or *magnetic variable). The distance from the Sun is 62 light-years.

Corona The outermost part of the Sun's atmosphere; it is made up of very tenuous gas at a very high temperature, and it is of great extent. It is visible with the naked eye only during a total solar *eclipse. Using a special instrument known as a *coronagraph, invented by the French astronomer B. *Lyot, the inner parts may be studied without an eclipse, but such observations are far from easy. Our knowledge of the corona has been vastly increased by observations made from space vehicles.

Corona Australid Meteors A minor southern shower, reaching its maximum in mid-March. It is not rich; the usual *ZHR is only about 5.

Total eclipse of the Sun (R. Maddison).

The Crab Nebula.

Coronagraph This is described under the heading *Corona.

Coronal hole A region of very low density and lower temperature in the Sun's *corona. Electrified particles can escape through these holes to produce the *solar wind.

Cosmic rays High-speed particles reaching the Earth from outer space. Some of them come from the Sun (and from Jupiter), but most originate far beyond the Solar System. The heaviest cosmic-ray particles are broken up when they dash into the upper air, so that at ground level we are shielded from them.

Cosmic year The name often given to the Sun's revolution period round the centre of the *Galaxy: about 225,000,000 years.

Cosmogony The branch of science dealing with the origin and development of the *universe, or any particular part of it.

Cosmological red shift The effect of the expansion of the universe in producing the *red shifts in the spectra of *galaxies.

Cosmology The study of the *universe as a whole; its nature, and the relations between its various parts.

Cosmos satellites Earth satellites launched by the Russians, beginning in March 1962. They are of many types, some scientific and some military. Well over a thousand have now been sent up.

Coudé system An optical system in which the light from the body under observation is received in a fixed direction (see also *Cœlostat). This means that the apparatus used to analyze the light need not be moved—a great advantage when the equipment is heavy. Most major reflectors can now be used with a Coudé focus.

Counterglow The English name for the sky-glow more commonly known by its German title, the *Gegenschein.

Crab Nebula (Messier 1: NGC 1952) A

cloud of expanding gas in Taurus, near the third-magnitude star Zeta Tauri. It is much too faint to be seen with the naked eye; good binoculars will show it, and it is an easy telescopic object. Photographs taken with large telescopes reveal a very complex structure.

The Crab Nebula is 6,000 light-years away. It is one of the strongest radio sources in the sky; it is also a source of X-rays and gamma-rays. It contains a *pulsar, which has been visually identified with a faint, flashing object with a period of a third of a second. It is the remnant of the *supernova of 1054, which became bright enough to be visible with the naked eye in broad daylight. In our Galaxy, the Crab is unique in our experience, and it is certainly one of the most informative objects in the entire sky.

Crêpe Ring The usual name for Saturn's Ring C; it is also known as the Dusky Ring.

The Cygnus Loop.

Crimean Astrophysical Observatory One of the main Russian observatories; it includes a 102-in reflector.

Crisium, Mare The Sea of Crises; a well-formed lunar sea, separate from the main system and easily visible with the naked eye not far from the north-east limb. It contains few craters; the two most prominent are Picard and Peirce. The brilliant ray-crater Proclus lies outside its western boundary.

Crommelin, Andrew Claude de la Cherois (1865-1939) English astronomer, noted for his work on cometary orbits. The most accurate predictions for the 1910 return of *Halley's Comet were made by Crommelin and his colleague Cowell.

Crommelin's Comet A medium-period comet. It was seen at different returns by four astronomers—Pons, Coggia, Winnecke and Forbes—and it was A. C. D. *Crommelin who realized that these

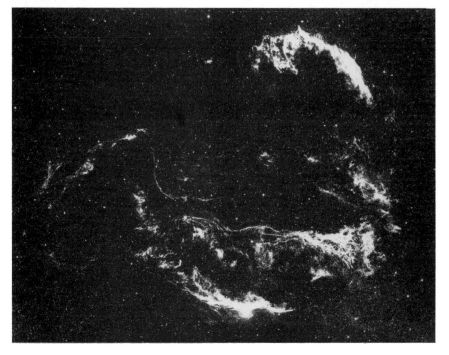

comets were identical. The last return was that of 1984, when the comet was used as a 'rehearsal' for Halley. It is not bright enough to be seen with the naked eye.

Crux Australis The *Southern Cross, the most famous of all the southern constellations. It is actually the smallest constellation in the sky, but is also exceptionally rich. In shape it is more like a kite than a cross!

Culmination The maximum altitude of a celestial body above the horizon. The Sun, of course, culminates at noon.

Cygnus A A very strong radio source in Cygnus; it is a large galaxy, but its immense distance (about 1,000 million light-years) means that it is optically very faint.

Cygnus Loop An *emission nebula in Cygnus; the *Veil Nebula forms part of it. It is an old *supernova remnant, but as the outburst occurred about 50,000 years ago there are no records of it.

Cygnus X-1 The system containing the star HDE 226868 and a possible Black Hole. See the heading *Black Hole.

D

D lines Bright yellow lines in the spectrum of sodium, seen as absorption lines in the spectrum of the Sun.

Dark nebula A 'cloud' of dust and gas, blotting out the light of objects beyond. The most famous example is the *Coal Sack in Crux. There is no essential difference between a bright nebula and a dark one; it all depends upon the presence, or otherwise, of suitable stars to provide illumination.

D'Arrest, Heinrich Ludwig (1822-1875) German astronomer who worked first at the Berlin Observatory (where he collaborated with *Galle in the discovery of Neptune) and then at Copenhagen Observatory. He concentrated upon comets and asteroids, but also discovered many nebulæ.

D'Arrest's Comet A periodical comet discovered by H. L. *D'Arrest in 1851. It has a period of 6.2 years, and has been seen regularly; at its best it may become just visible with the naked eye.

Darwin, Sir George (1845-1912) Son of Charles Darwin. He is best remembered for his tidal theory of the origin of the Moon.

Davida *Minor planet No 511. It has a diameter of 200 miles, and is one of the largest members of the asteroid swarm. Its mean opposition magnitude is 10.5; the period is 5.7 years.

Dawes limit The practical limit for the *resolving power of a telescope; it is $4.56/d$, where d is the aperture of the telescope in inches and the resolution is in *seconds of arc.

Day The period taken for the Earth to spin once on its axis. However, there are various kinds of 'days'.

A *sidereal day* is the time interval between successive meridian passages, or *culminations, of the same star (23 hours 56 minutes 4.091 seconds). In other words, the Earth's rotation period is measured with respect to the stars, which are so far away that for all practical purposes we may regard them as infinitely remote.

A *solar day* is the time interval between two successive noons. It is slightly longer than the sidereal day, since the Sun is moving against the starry background at about one degree a day in an easterly direction. The situation is complicated by the fact that the Earth's orbit is somewhat elliptical; when at its nearest to the Sun, the Earth moves fastest, according to the principles of *Kepler's Laws, and so the Sun seems to move at its fastest against the stars. (The stars cannot be seen in daylight only because they are overpowered by the brilliance of the Sun and the sky.) For convenience, astronomers normally measure by a *mean sun*, which is an imaginary body moving round the celestial equator at a constant speed which is equal to the average rate of motion of the true Sun around the *ecliptic. The *mean solar day* is 24 hours 3 minutes 56.555 seconds long.

Astronomically, the 24-hour clock is used, midnight being 0 hours. (In a *civil day* there are two 12-hour periods, am and pm.) For scientific purposes, daylight-saving adjustments such as British Summer Time are ignored; everything is reckoned according to *Greenwich Mean Time.

Declination The angular distance of a celestial body north or south of the celestial *equator. Therefore, the equator itself has a declination of 0°, while the north celestial pole is at 90°N (+90°) and the south celestial pole 90°S (−90°). Declination in the sky corresponds to latitude on the Earth.

Deferent In *Ptolemy's system of the universe, all the bodies in the sky moved in circular orbits; the circle was the 'perfect' form, and nothing short of perfection could be allowed in the heavens! Unfortunately, it was obvious that the movements of the planets could not be accounted for on the theory that they moved round the Earth at constant speeds in perfectly circular paths. Ptolemy, who was an excellent observer and mathema-

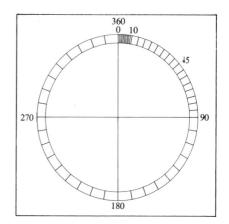

Degrees in a circle.

tician, knew this quite well, so he overcame the difficulty by assuming that a planet moved round the Earth in a small circle or *epicycle, the centre of which—the deferent—itself moved round the Earth in a perfect circle. As more and more irregularities came to hand, more and more epicycles had to be introduced, until the whole system became hopelessly clumsy and artificial. All the same, it lasted for many centuries, and it was only in the years following 1543 that it was replaced by the *Copernican system.

Degenerate matter Matter in which all the electrons have been stripped away from

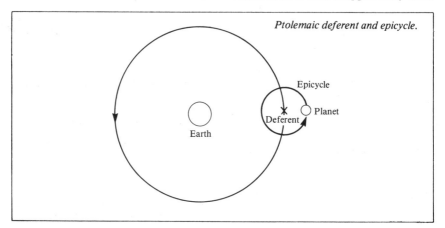

Ptolemaic deferent and epicycle.

their nuclei, so that the matter can become remarkably dense. With increased density, the electrons become so numerous over a given area that they can exert great pressure (*degeneracy pressure*) which can prevent further compression in a star of no more than 1.44 times the mass of the Sun, which is the limiting upper value for a *White Dwarf. If the mass is still greater, the collapsing star becomes a *neutron star.

Degree of arc A unit for measuring angles. A full circle contains 360 degrees; therefore a right angle contains one quarter of 360 degrees = 90 degrees. Each degree is subdivided into 60 minutes of arc, and each minute is again divided into 60 seconds of arc. The symbols are ° (degree), ' (minute) and " (second). One second of arc is the apparent length of a foot rule at a distance of 32 miles!

Deimos The outer and smaller satellite of *Mars. For data, see *Satellites. It is difficult to see with small telescopes owing to its small size and its closeness to Mars; it was discovered by Asaph *Hall in 1877.

Pictures obtained from the *Mariner 9 and *Viking space-craft show that Deimos is irregularly shaped, with a longest diameter of no more than 10 miles, and that it is cratered, though to a lesser extent than the other Martian satellite, *Phobos; its surface presumably has a deeper *regolith. The two largest craters are named Swift and Voltaire. Seen from Mars, Deimos would have an apparent diameter of only about 2 minutes of arc. It would remain above the Martian horizon for 2½ Martian days or *sols consecutively, but from latitudes higher than 82°N or 82°S it would never rise above the horizon. It would transit the Sun about 130 times in each Martian year. It is possible that both Deimos and Phobos are captured asteroids rather than bona-fide satellites.

De la Rue, Warren (1815-1889) Channel Islands amateur astronomer. He was a pioneer of astronomical photography.

Delta Aquarid Meteors A minor shower, reaching its maximum about 28 July. The *ZHR can rise to 35.

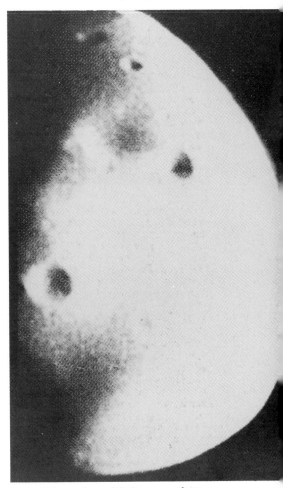

Part of the surface of Deimos seen from a Viking probe.

Delta Cephei The prototype Cepheid. It has a period of 5.3 days and a magnitude range of from 3.5 to 4.4, so that it is always easily visible with the naked eye.

Delta Scuti variables Pulsating *variable stars, with short periods (of a few hours) and very small magnitude ranges. Most of them are of spectral types A or F.

Deneb The star Alpha Cygni; for data, see *Stars. It is one of the most remote and

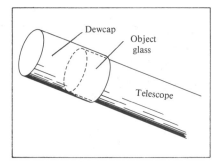

Dewcap.

most luminous of the first-magnitude stars; we see it today as it used to be at the time of the Roman Occupation!

Denebola The star Beta Leonis. Its magnitude is 2.14, but in ancient times it was ranked as a first-magnitude star. Any permanent change seems unlikely, since the spectral type is A3, though it is true that it has been suspected of slight variability.

Denning, William Frederick (1848-1931) English amateur astronomer, who discovered several comets and made useful planetary observations. His main work was in connection with *meteors; he determined the positions of many radiants.

Density The amount of matter in a unit volume of substance. For most purposes, water is taken as 1; thus the mean density of the Earth (the *specific gravity*) is 5.52, ie, a given volume of the Earth 'weighs' 5.52 times as much as an equal volume of water would do. (It must be remembered that this relates to the Earth as a whole. Near the outer crust, the density is well below 5; near the core, it rises to 9 or 10.) Taking water as unity, the Sun's mean density is only 1.4, and that of Saturn is less than 1. With stars such as *White Dwarfs and *neutron stars, the densities are remarkably high.

Descartes, René (1596-1650) French philosopher and mathematician, who developed the theory that matter origi-

nates in the so-called æther which fills all space. He also made improvements to telescopic lenses.

Descending node See *Nodes.

De Sitter, Willem (1872-1935) Dutch astronomer; a leading cosmologist. He was Director of the Leiden Observatory from 1919 to 1935.

Deslandres, Henri Alexandre (1853-1948) French astronomer; in 1891 he invented the *spectroheliograph, independently of *Hale. In 1908 he became Director of the *Meudon Observatory.

Deuterium 'Heavy *hydrogen'; the nucleus consists of one proton and one neutron. Stars do not create deuterium; they destroy it, and so it was widely believed that there should be none of it in space. However, it has now been found that there is a great deal of deuterium, which seems to indicate that element-building must have occurred in places other than the interiors of stars—presumably at the time of the origin of the universe with the *Big Bang.

Diamond Ring effect.

Dewcap An open tube fitted to the end of a refracting telescope, beyond the *object-glass. It prevents the object-glass from becoming wet by condensation—at least, it should do so!

Diagonal In a *Newtonian refractor, a small flat mirror placed in the upper tube at an angle of 45°, to direct the light-rays to a focus in the side of the tube. It is often called simply a *flat*.

Diamond ring effect The appearance just before the beginning and just after the end of totality at an *eclipse of the Sun, when a very small part of the brilliant solar disk shines out beyond the edge of the dark disk of the Moon.

Dichotomy The exact half-*phase of Mercury, Venus or the Moon.

Differential rotation The type of rotation shown by a body which is not solid, such as the Sun or a giant planet. With *Jupiter, for instance, the mean rotation of the equatorial zone is 9 hours 50 minutes 30 seconds, while that of the rest of the planet is 9 hours 55 minutes 41 seconds.

Diffraction grating A device used for splitting up light; it may be regarded as an alternative to the *prism of a *spectroscope. The grating consists of a series of close parallel lines ruled on a polished metallic surface. The lines must be very close indeed—several thousands per inch —so that diffraction gratings are not easy to make.

Diffraction rings Concentric rings surrounding the image of a star as seen in a telescope. They cannot be eliminated, since they are due to the wave-nature of light and also the construction of the telescope; however, they are more of a nuisance in small telescopes than in large ones.

Dione The fourth satellite of *Saturn. For data, see *Satellites. It is a medium-sized icy satellite, slightly larger than *Tethys but appreciably smaller than *Rhea or *Iapetus. It has a density of about 1.4 times that of water—greater than for the

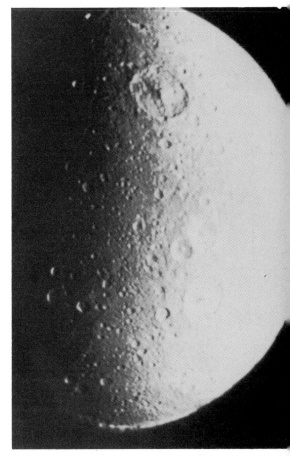

Dione, photographed from a Voyager probe.

other icy satellites—and it may play a rôle in the flexing of the interior of *Enceladus, whose revolution period is almost exactly half that of Dione. The *Voyager pictures show that the surface is not uniform, with a darkish trailing hemisphere and a brighter leading hemisphere (the rotation is, of course, *captured). The most prominent feature, Amata, is 150 miles in diameter; it may be either a crater or a basin. There are two very prominent craters, Aeneas and Dido. A system of bright wispy features extends over the darker hemisphere, due possibly to ice which has seeped out from below. A small

47

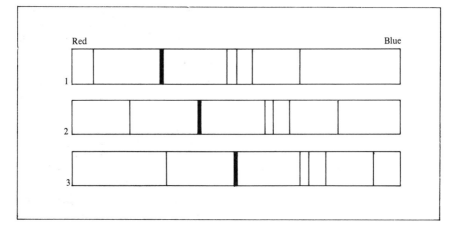

The Doppler effect: **1** *Receding;* **2** *Stationary;* **3** *Approaching.*

satellite, as yet unnamed, moves in the same orbit as Dione.

Dipper American unofficial name for the *Plough in *Ursa Major (the Great Bear).

Direct motion Bodies which move round the Sun in the same sense as the Earth are said to have direct motion; those which move in the opposite sense have *retrograde motion. The terms may also be used with respect to satellites. No planet or asteroid with retrograde motion is known, but there are various retrograde comets (including *Halley's) and satellites.

The terms are also used with regard to the apparent movements of the planets in the sky. When shifting eastward against the stars, a planet is moving in the direct sense; when moving westwards, it is retrograding.

Distance modulus The difference between the absolute magnitude (M) and the apparent magnitude (m) of a star; M-m is a measure of the distance of the star. The principle is of great value in estimating the distances of very remote objects.

Diurnal motion The apparent daily rotation of the sky from east to west, due to the real rotation of the Earth from west to east.

Dog Star A nickname for *Sirius, the brightest star in the sky.

Donati's Comet A brilliant comet seen in 1858, discovered by the Italian astronomer G. Donati (a pioneer in studies of cometary spectra). With its scimitar-like tail, it is said to have been the most beautiful comet ever observed. The period is thought to be about 2,000 years.

Doppler Effect The apparent change in wavelength of light (or sound) caused by the motion of the source or of the observer.

The classic example is that of a whistling train by-passing an observer. When the train is approaching, more sound-waves per second enter the observer's ear than would be the case if the train were standing still; the wavelength is effectively shortened, and the whistle is high-pitched. When the train starts to recede, fewer sound-waves per second reach the ear; the wavelength is lengthened, and the note of the whistle drops. It is much the same with light, but here the effect is to make the object 'too blue' when approaching and 'too red' when receding.

The actual changes in colour are slight, but the effects show up in the spectra of the sources. If the object is approaching, the spectral lines are shifted over to the short-wave or blue end of the spectrum; if the object is receding, there is a *red shift.

All the external *galaxies, except those of our *Local Group, are receding, and show red shifts; the greater the distance, the greater the velocity of recession—and the greater the red shift.

The Doppler Effect is named in honour of the Austrian physicist who first drew attention to it, in 1842. It is of fundamental importance in astronomical studies.

Double star A star made up of two components relatively close together in the sky. Some doubles are *optical*; that is to say, the components are not genuinely associated, but simply happen to lie near the same line of sight as seen from Earth. Most, however, are true *binary systems.

Draconids See *Giacobinids.

Draper, Henry (1837-1882) American pioneer of astronomical photography.

Draper Catalogue An important catalogue of stellar spectra, due mainly to Annie *Cannon and published in 1924. It was financed by money provided by Henry *Draper's widow.

Dreyer, J. L. E. (1852-1926) Danish astronomer. He spent many years in Ireland, first with Lord *Rosse and then as Director of the *Armagh Observatory, before retiring to Oxford. He compiled the NGC or New General Catalogue of clusters and nebulæ, in 1888, and was also a noted astronomical historian.

Driving clock A mechanical device used to move a telescope round just fast enough to compensate for the Earth's rotation. Electrical drives are now used, but are still often referred to as clocks.

Dubhe The star Alpha Ursæ Majoris; the brighter of the two Pointers to Polaris. For data, see *Stars. It and *Alkaid are the only two stars in the Plough pattern which do not belong to the Ursa Major *moving cluster.

Dumbbell Nebula Messier 27, a *planetary nebula in Vulpecula. It is about 3° north of the star Gamma Sagittæ. It is about 975 light-years away. The characteristic dumbbell form can be seen with a fairly small telescope, and M.27 is usually regarded as the finest of all planetary nebulæ.

Dunsink Observatory The main observatory in Eire, founded in 1785. It is five miles from Dublin.

Dust rings in the Solar System Dust rings discovered in 1983 from the infra-red satellite *IRAS. They are well away from the main plane of the system, and are at about the distance of the asteroid zone.

Dwarf Novæ A common term for *SS Cygni (or U Geminorum) variables.

Dyson, Sir Frank (1868-1939) The ninth Astronomer Royal; his term of office extended between 1910 and 1933.

E

Eagle Nebula Messier 16, a gaseous *nebula and associated galactic cluster in the constellation of Serpens. It is over 5,500 light-years away. It contains many *Bok globules. It was discovered by the French astronomer de Chéseaux in 1746; it is a fine telescopic object, though the nebulosity is rather faint.

Early type stars Stars of spectral type W, O, B and A. The name was given when it was still believed that the spectral sequence was a true evolutionary sequence.

Earth The third planet in order of distance from the Sun. Data are given under the heading *Planets.

The mean distance between the Sun and the Earth is 92,957,209 miles, but since the orbit is not a perfect circle the distance ranges between 91,400,000 miles at *perihelion out to 94,600,000 miles at *aphelion. The *seasons are due not to this changing distance, but to the fact that the Earth's axis is inclined to the perpendicular to the orbital plane by an angle of 23½°.

The Earth is the largest of the four inner planets, though Venus is almost its equal. Its mass is some 6,000,000,000,000,000,000,000 tons, and its mean *density or specific gravity is 5.52 times that of water. There is a heavy, iron-rich core. The *atmosphere is made up chiefly of nitrogen (77.6 per cent) and oxygen (20.7 per cent); no other planet or satellite in the Solar System has nearly so much free atmospheric oxygen, so that there is nowhere else where an Earth-type man or animal could breathe.

In shape, the Earth is not a perfect sphere. Its diameter is 7,926 miles as measured through the equator, but only 7,900 miles as measured through the poles. This is because the axial rotation is fairly rapid, causing the equatorial regions to 'bulge' slightly.

The age of the Earth is about 4.7 thousand million years. Like the other planets, the Earth was formed from the so-called solar nebula by the *accretion process.

Earth-grazers Nickname for the *minor planets which may make close approaches to the Earth (see *Amor, *Apollo, *Aten).

Earthshine When the Moon appears as a crescent, the 'dark' side may often be seen shining dimly. This is because the night side of the Moon is being lit up by light reflected from the Earth.

Easter A religious festival, the date of which is determined by the Moon. According to a decree by the Council of Nicæa in 325, Easter falls on the first Sunday after the first full moon after the *vernal equinox.

Echelle grating A *diffraction grating with very fine lines, ruled wider apart than with other gratings. It gives high resolution over a narrow range of wavelengths.

Echo satellites Two balloon satellites, launched from the United States in the 1960s; they were passive reflectors. Both were bright, and were widely observed. Eventually they decayed in the Earth's atmosphere.

Eclipses, Lunar Because the Moon revolves round the Earth (or, to be more accurate, because both bodies move round their common centre of gravity, or *barycentre), there must be times when the Moon passes into the shadow cast by the Earth. This causes a lunar eclipse.

During an eclipse, the Moon does not (usually) vanish completely; some sunlight is bent or refracted on to its surface by the Earth's atmosphere, so that the Moon merely turns a dim, often coppery colour until it passes out of the shadow again. Lunar eclipses may be either total, when the whole of the Moon passes into the shadow, or partial, when only a portion of the Moon is covered. Obviously, a lunar eclipse can happen only at full moon; the reason why an eclipse is not seen every month is because the Moon's

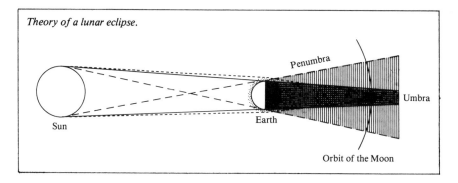

Theory of a lunar eclipse.

Penumbra

Umbra

Sun

Earth

Orbit of the Moon

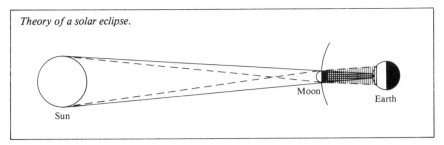

Theory of a solar eclipse.

Sun

Moon

Earth

orbit is tilted relative to that of the Earth, and no eclipse can take place unless the full moon is close to a *node. Totality can never last for more than 1¾ hours.

Since the Sun is a disk, and not a point source of light, there is an area of partial shadow or *penumbra* to either side of the main cone of shadow or *umbra* cast by the Earth. If the Moon misses the main shadow, but enters the penumbra, a *penumbral eclipse* results, and the slight dimming may just be noticed with the naked eye. Of course, the Moon must always pass through the penumbra before entering the main cone.

As soon as the sunlight is cut off from the Moon, the lunar surface temperature drops sharply; it may fall by over 200° Fahrenheit in a single hour, showing that the surface materials are very poor at retaining heat. On the other hand, there are some areas on the Moon—notably the prominent 54-mile ray-crater *Tycho— which cool down more slowly, and are known, rather misleadingly, as 'hot spots'. Infra-red studies of them have been most informative.

Eclipses of the Moon are not of great importance astronomically, but they are always worth watching, since beautiful coloured glows on the Moon may some- times be seen; they are, of course, due to the light which has passed through the Earth's air, and have nothing directly to do with the Moon itself.

If a lunar eclipse occurs, it is visible over an entire hemisphere of the Earth. Therefore, any particular place on Earth will see eclipses of the Moon more frequently than those of the Sun. The following lunar eclipses will take place between 1986 and 1990:

1986	24 April	Total
1986	17 October	Total
1987	7 October	Partial: 1 per cent eclipsed.
1988	27 August	Partial: 30 per cent eclipsed.
1989	20 February	Total
1989	17 August	Total
1990	9 February	Total
	6 August	Partial: 68 per cent eclipsed.

Eclipses, Solar By a lucky chance the Sun and the Moon appear almost the same size in the sky. Therefore, the Moon may sometimes pass in front of the Sun, hiding or eclipsing it. Strictly speaking, what we call a solar eclipse is really an *occultation of the Sun by the Moon.

Solar eclipses are of three kinds. In a *total eclipse*, the Sun is completely hidden, and the effect is magnificent, since the *corona, *chromosphere and *prominences flash into view. If the eclipse is *partial*, none of these pheno- mena can be seen with the naked eye. If exact lining-up occurs when the Moon is near its greatest distance from the Earth, the lunar disk appears smaller than that of the Sun, and cannot wholly cover it, so that a ring of sunlight is left showing round the dark disk of the Moon, pro- ducing an *annular* eclipse.

Obviously, a solar eclipse can happen only at new moon. The 5°9' tilt of the Moon's orbit with respect to that of the Earth means that eclipses do not happen every month; usually the new moon passes unseen either above or below the Sun in the sky. Moreover, a solar eclipse is not visible over an entire hemisphere of the Earth, as with an eclipse of the Moon.

Total eclipses are of great importance to astronomers, since there are various investigations which cannot be carried out by Earth-based observers at any other time. Unfortunately they are not so common as might be wished, and the track of totality can never be more than 169 miles wide; totality can never last for as long as 8 minutes, and is generally much less than this. The last total eclipse visible from anywhere in England was that of 1927; the next will be on 11 August 1999, when the track will cross Cornwall.

The following solar eclipses will take place between 1986 and 1990:

1986	9 April	Partial: 82 per cent eclipsed. Antarctic area.
1986	3 October	Total, but annular along most of the track. Atlantic area.
1987	29 March	Total, but annular along most of the track. Argentina to Indian Ocean.
1987	23 September	Annular. Russia, China and Pacific area.
1988	18 March	Total. Indian Ocean to the Pacific area.
1989	7 March	Partial: 83 per cent eclipsed. Arctic area.
1989	31 August	Partial: 63 per cent eclipsed. Antarctic area.
1990	26 January	Annular. Antarctic area.
1990	22 July	Total. Finland, Russia, Pacific area.

Eclipsing binary See *Eclipsing variable.

Eclipsing variable (or eclipsing binary) A *binary system, made up of two components moving round their common centre of gravity in such a way that one star passes periodically in front of the other as seen from the Earth, so cutting out part of its light. The best-known member of the class is *Algol in Perseus. Other types are composed of components which are less unequal, so that there are two minima, one deeper than the other; the prototype star is Beta Lyræ. With *W Ursæ Majoris pairs, both components are dwarfs and are almost in contact. Eclipsing variables are most useful, because their orbits can be well determined—and this leads on to an evaluation of the total mass of the system.

Ecliptic The projection of the Earth's orbit on to the *celestial sphere. It may

Eclipsing binary.

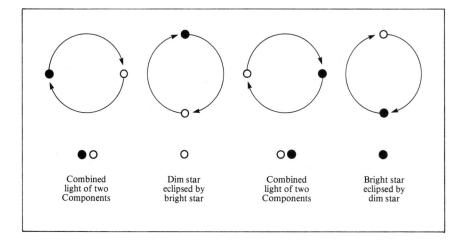

| Combined light of two Components | Dim star eclipsed by bright star | Combined light of two Components | Bright star eclipsed by dim star |

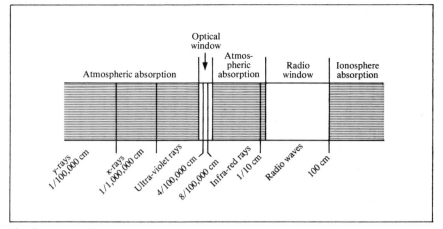

The electromagnetic spectrum.

also be defined as the apparent yearly path of the Sun against the stars, passing through the twelve constellations of the *Zodiac. Since the plane of the Earth's orbit is inclined to the equator by 23½°, the angle between the ecliptic and the celestial equator is also 23½°.

Ecosphere The region round a star where, other conditions being suitable, the temperature is neither too high nor too low for Earth-type life to exist. In the Solar System, the Earth lies in the middle of the ecosphere, with Mars at the outer edge and Venus at the inner edge.

Eddington, Sir Arthur Stanley (1882-1944) English astronomer; one of the greatest of all cosmologists, who also undertook pioneer work into the constitution and evolution of the stars.

Effective temperature The surface temperature of a star (or other body) expressed as the temperature of a *black body having the same radius and emitting the same total amount of energy.

Einstein, Albert (1879-1955) German physicist who laid down the principles of *relativity theory; probably the greatest of all mathematicians since *Newton. He lived in America from 1933 until his death.

Einstein Observatory An artificial satellite, launched on 13 November 1978 from Cape *Canaveral. It was concerned with studying X-rays from space, and was very successful; it continued to operate until early 1981. It was 21 ft long, and orbited the Earth at a height of 500 miles above the ground.

Elara The seventh satellite of Jupiter. For data, see *Satellites.

Electromagnetic spectrum The full range of wavelengths, from radio waves through to gamma-rays. Visible light occupies only a very small part of the total electromagnetic spectrum.

Electron A fundamental particle carrying unit negative charge; it makes up part of an *atom. Each electron is almost inconceivably small, and the number needed to make a weight of one ounce has been given as 311×10^{26}—in other words, 311 followed by 26 zeros. However, it is misleading to picture an electron as being a solid lump.

Electron density The number of free electrons in unit volume of space. (A free electron is an electron which is not attached to any particular atom, but is moving around on its own.)

Electronic aids Modern devices, used with large telescopes, which by now have largely superseded *photography for most branches of research.

Element A substance which cannot be chemically split up into simpler substances. It may be said that the elements are the fundamental 'building blocks' of the universe. Familiar elements include hydrogen, helium, oxygen, iron, gold, silver, mercury and tin; all other substances are made up of combinations of elements. Ninety-two elements are known to occur naturally, the lightest being hydrogen and the heaviest uranium.

Elongation The apparent angular distance of a planet or comet from the Sun, or of a satellite from its primary planet. When a planet is at *opposition, and so is directly opposite to the Sun in the sky, its elongation is 180°. This can never apply to Mercury or Venus, which are closer to the Sun than we are, and are always to be found somewhere near the Sun in the sky; the maximum elongation is 47° for Venus, only 28° for Mercury.

Elysium Planitia A plain in the southern hemisphere of Mars. It is a volcanic region, with one particularly lofty volcano, Elysium Mons, rising to a height of over 9 miles.

Emission spectrum A spectrum consisting of isolated bright lines, each of which is characteristic of a particular *element or group of elements.

Enceladus The second satellite of Saturn. For data, see *Satellites.

The density of Enceladus is not much greater than that of water, and its *albedo is virtually 100 per cent, making it the most reflective body in the Solar System. *Voyager pictures show that although craters exist, most of them are small and young-looking, while there are some crater-free areas dominated by long grooves. In spite of its small size, Enceladus may be an active world, possibly because its interior is being 'flexed' by the gravitational pull of the much more massive satellite *Dione, in which case soft ice or even water may well up from below the crust, obliterating existing surface features there. Moreover, Enceladus is exceptionally cold even by Saturnian standards. The average day temperature is only about −200° Centigrade.

Encke, Johann Franz (1791-1865) German astronomer, and Director of the Berlin Observatory. He calculated the distance of the Sun from observations of the *transits of Venus of 1761 and 1769, and discovered the periodicity of the *comet which now bears his name. He also authorized the search for Neptune carried out by two astronomers at the Observatory, *Galle and *D'Arrest.

Encke's Comet The comet with the shortest known period (3.3 years). Its distance from the Sun ranges between 0.3 and 4.1 *astronomical units, and with modern equipment it can be followed throughout its orbit. It was discovered by P. Méchain in 1786, and seen again by Caroline *Herschel in 1795, when it was just visible with the naked eye. The next sighting was by *Pons in 1805, and Pons again discovered it in 1818. The orbit was computed by J. F. *Encke, who successfully predicted its return in 1822. It was fitting that the comet should be named in Encke's honour; up to that time the only comet known to be periodical was *Halley's.

Encke's Division A gap in Saturn's Ring A, discovered by J. F. *Encke. It is not a difficult telescopic object when the ring-system is suitably placed.

Ephemeris A table showing the predicted positions of a moving celestial body, such as a comet or a planet.

Epicycle This is described under the heading *Deferent.

Epidemiarum, Palus The 'Marsh of Epidemics', a minor lunar sea leading off the Mare *Nubium.

Epimetheus A satellite of Saturn. For data, see *Satellites. It is co-orbital with another small satellite, *Janus.

Epoch A date chosen for reference purposes in quoting astronomical data.

Epsilon Aurigæ One of the most remarkable stars in the sky. It lies near Capella, and has a normal magnitude of 3.0, but every 27 years it fades by about a magnitude due to eclipse by an invisible companion. The eclipse is protracted; totality lasts for a year. The visible star is a very luminous F-type supergiant, with at least 60,000 times the power of the Sun and over 4,500 light-years away; the companion, once regarded as a possible *Black Hole, is now thought to be a hot bluish star surrounded by an extensive, opaque cloud of material. Epsilon Aurigæ is the longest-period eclipsing binary known; the last eclipse ended in 1984. It is one of a triangle of stars known as the Haedi or 'Kids'; the other two are Eta Aurigæ, a normal B-type star, and the eclipsing binary *Zeta Aurigæ.

Epsilon Eridani A K-type dwarf star, apparent magnitude 3.8; it has only one-third the luminosity of the Sun. Its distance is 10.7 light-years. It is one of the nearest stars which is at all comparable with the Sun, and it has been regarded as a possible centre of a planetary system; during Project *Ozma, efforts were made

Ecliptic and celestial equator.

to detect artificial signals from it, though, hardly surprisingly, the results were negative.

Epsilon Indi One of the nearest stars, at a distance of 11.6 light-years. Its apparent magnitude is 4.7; it is a K5-type dwarf, with a luminosity of only 17 per cent that of the Sun.

Epsilon Lyræ A multiple star, near *Vega. The combined magnitude is 3.9. The two components are separated by 207.8″, so that keen-eyed people can distinguish them without optical aid; telescopically each component is seen to be again double, making up a quadruple system. The two main pairs are about a fifth of a light-year apart.

Equation of time As explained under the heading *Day, the Sun does not move against the stars at a constant speed, because the Earth's orbit is not a perfect circle, and astronomers make use of a *mean sun*, which travels along the celestial equator at a speed equal to the average speed of the real Sun. The interval by which the real Sun is ahead of or behind the imaginary mean sun is termed the equation of time. It can never be greater than 17 minutes; four times every year it becomes zero, so that the right ascension

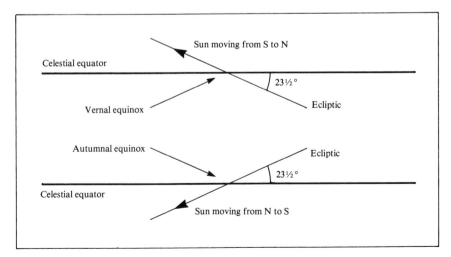

of the mean sun is then the same as the right ascension of the real Sun.

Equator, Celestial The projection of the Earth's equator on to the *celestial sphere. It divides the sky into two equal hemispheres. Delta Orionis or *Mintaka, the northernmost of the three stars in Orion's Belt, lies very close to the celestial equator, so that its *declination is practically 0°. Another naked-eye star close to the equator is Zeta Virginis, magnitude 3.4.

Equatorial mount If a telescope is mounted upon an axis which is parallel to the axis of the Earth, it need be moved only in *right ascension (east to west) to keep a celestial body in the field of view; the up-or-down movement (*declination) will look after itself. The polar axis, on which the telescope is mounted, points to the celestial pole. There are various types of equatorial mounts, but all depend upon the same principle. With the *German* type, a second shaft extends at right angles from the polar axis; one side takes the telescope, while the other side carries a counterweight. In the *English* type, the

telescope is pivoted inside a large yoke, inclined at the correct angle and supported by piers. Better, perhaps, is the *Fork*, which is not unlike the English but has no pier; the upper part of the yoke is missing, and the telescope tube is pivoted between the prongs of the fork. With the *Foucault*, the polar axis broadens out into a large disk; two stout arms are fixed to the face of the disk, forming the arms between which the telescope is mounted.

Equatorials make driving the telescope much easier than with an *altazimuth, since only one motion has to be allowed for. Until recently all large telescopes were equatorial, though by now computerized altazimuth mountings are becoming popular.

Equinox Twice a year the Sun crosses the celestial *equator, once when moving from south to north (about 21 March) and once when moving from north to south (about 22 September). These points are the two equinoxes. That of March is called the spring or *vernal equinox (First Point of Aries), while that of September is the autumnal equinox (First Point of Libra). Another way of expressing this is to say that the equinoxes are the two points where the *ecliptic cuts the celestial

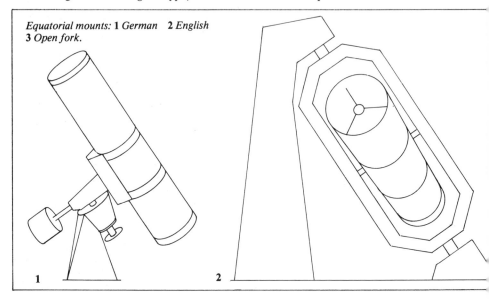

Equatorial mounts: **1** *German* **2** *English* **3** *Open fork.*

1

2

equator. Owing to *precession, the First Point of Aries is now in Pisces, while the First Point of Libra has moved into Virgo.

Eratosthenes (BC 276-196) Greek astronomer, born in Cyrene and for many years Librarian at Alexandria. He measured the size of the Earth with remarkable accuracy.

Ergosphere The region immediately outside the *event horizon of a rotating *Black Hole.

Eros A small *minor planet; its longest diameter is less than 20 miles, and its maximum width is about 9 miles. Though its mean distance from the Sun is 135,600,000 miles, it has an eccentric orbit which sometimes brings it within 15,000,000 miles of the Earth, and during the close approach of 1931 it was very carefully studied, because measures of its position allowed its orbit to be worked out very accurately—which, in turn, provided a key to the length of the *astronomical unit or Earth-Sun distance. Better ways of measuring the astronomical unit are now used, so that the last approach of Eros, in 1974-5, was not regarded as important.

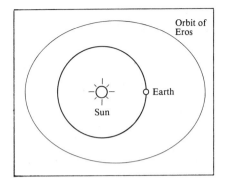

Orbit of Eros.

Eros rotates in a period of 5 hours 16 minutes, and this, of course, causes changes in its brightness; it is most brilliant when 'broadside-on' to us. Telescopically it looks like a star, though in 1931 van den Bos and his colleagues in South Africa, using the 27-in Johannesburg refractor, were able to see its elongated shape.

Eros was discovered in 1898, photographically by Witt in Berlin; it was photographed on the same night by Charlois, in France, who however was less quick at comparing his plates and so missed the honour of being co-discoverer. Eros is No 433 in the list of minor planets. Since it was the first known minor planet to come within the orbit of Mars, it was also the first to be given a masculine name.

Escape velocity The minimum velocity at which an object must move in order to escape from the surface of a planet, or other body, without being given any extra impetus. It is 7 miles per second in the case of the Earth.

Eta Aquilæ With Delta Cephei, the brightest of the *Cepheid variables. Its fluctuations were discovered by Pigott in 1784. It has a period of 7.2 days, and a magnitude range of between 3.4 and 4.7.

Eta Carinæ The most erratic of all *variable stars. At one time, in the 1830s and early 1840s, it shone as the brightest star in the sky apart from *Sirius, and

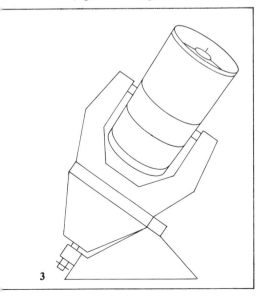

reached magnitude −0.8, but throughout the present century it has been just below the limit of naked-eye visibility. Its distance is thought to be about 6,000 light-years, and at its peak it must have been 6,000,000 times as luminous as the Sun, making it the most powerful star known. It is still very luminous, but much of its energy is in the *infra-red, and telescopically it appears as an 'orange blob' quite unlike a normal star; it is associated with extensive nebulosity, and this no doubt cuts off some of its light (see also *Keyhole Nebula). It is exceptionally massive, with a very peculiar spectrum. Its precise nature is uncertain; there have been suggestions that it may be preparing to explode as a *supernova, though we cannot tell when this will happen! Eta Carinæ is unlike any other known object. It may brighten up again at any time, and European astronomers never cease to regret that it lies so far south in the sky.

Eudoxus (*c* BC 408-355) Greek astronomer who developed a theory of concentric spheres—the first truly scientific attempt to explain the movements of the planets.

Euler, L. (1707-1783) Swiss mathematician, who made outstanding contributions to dynamical astronomy, and continued his work even after he had lost his sight.

Europa The second *Galilean satellite of Jupiter. For data, see *Satellites.

Europa, with its diameter of 1,943 miles, is the only Galilean which is smaller than our Moon. It was studied from the *Voyager space-craft and found to be a most unusual world. There are virtually no craters, and the surface is icy; probably there is an ice crust at least 600 miles deep, overlying either an ocean of water or else a region of soft ice above a silicate core. There is little vertical relief; it has been said that Europa is as smooth as a billiard-ball. There are shallow grooves, very difficult to map, with low curved ridges and darker patches. If craters have ever existed on Europa, they have been obliterated, perhaps by water seeping out from below and then freezing. There have

even been suggestions that there might be life in the underground ocean, though this is, to say the least, highly speculative! It seems that the interior of Europa is 'churned' by a combination of gravitational flexing due to Jupiter, and the effects of the pull of *Io, but it is certainly hard to explain why Io and Europa are so different.

European Southern Observatory (ESO) A major observatory at La Silla, near La Sirena in Chile. The main telescope is a 152-in reflector. It is operated by observatories in Germany, Denmark, Sweden, Holland, Belgium and France.

Evection An inequality in the Moon's motion. It is due to the pull of the Sun, which affects the orbit of the Moon and makes it alternately a little more eccentric and a little less eccentric. Evection is predictable, and has to be taken into account when working out the times of phenomena such as *eclipses; the effect may amount to as much as 3 hours.

Exobiology The study of possible life beyond the Earth. So far it is theoretical only!

Exosphere The outermost part of the Earth's *atmosphere. It has no definite boundary, but simply thins out until its density is no greater than that of the interplanetary medium.

Explorer satellites American artificial satellites. The first, Explorer 1, was launched from Cape *Canaveral in June 1958 by a team led by Wernher *von Braun, and was the first US satellite; its instruments detected the *Van Allen radiation zones round the Earth. Over sixty Explorers have since been launched.

Extinction When a star or planet is low in the sky, its light comes to the observer through a relatively thick layer of atmosphere, so that the brightness is reduced. This effect is termed extinction. It amounts to three magnitudes for a star only 1° above the horizon, but to only one magnitude for a star at an altitude of 10°. Above an altitude of 45°, extinction is so

The surface of Jupiter showing both the Red Spot and two of the planet's satellites, Io, left, and Europa, right. Photograph taken by Voyager 1 in 1979 (JPL).

slight that for most practical purposes it may be neglected.

Extragalactic nebulæ An obsolete term for *galaxies.

Eyepiece (or ocular) The lens, or combination of lenses, placed at the eye-end of a telescope. Its role is to magnify the image formed by the object-glass of a refractor or the mirrors of a reflector; in fact all the magnification is done by the eyepiece. An astronomical telescope will have several eyepieces, so that different magnifications may be used as desired.

There are various types. With a *positive* eyepiece, the image-plane is outside the eyepiece—between it and the object-glass or mirror—so that it can be used with a *micrometer; examples are the Ramsden, Orthoscopic and Monocentric. With a *negative* eyepiece, such as the Huyghenian or Tolles, the image-plane lies inside the eyepiece. A *Barlow* is a concave lens of about 3 in negative focal length, mounted in a short tube which can be placed between the object-glass (or mirror) and the eyepiece, inside the drawtube of the eyepiece. It increases the effective focal length of the object-glass or mirror, and so provides extra magnification.

F

FU Orionis A very young star in the Orion nebulosity. In 1936 it brightened up from magnitude 16 to 10, and has since remained at that level. It is rich in lithium.

Fabricius, David (1564-1617) Dutch amateur astronomer; he was a minister in the Church, and was assassinated by one of its members. Fabricius was a pioneer telescopic observer, but is best remembered for having made the first recorded observation of *Mira Ceti, in 1596, though he did not recognize it as a variable star.

Faculæ Bright, temporary patches above the *photosphere of the Sun. They are usually associated with sunspots, and are easily visible with a small telescope (by projection, of course; direct observation of the Sun is very dangerous). Faculæ frequently appear in a position near which a spot-group is about to appear, and may persist for some time in the region of a group which has disappeared.

False colour technique The use of colours in a photographic image or a chart to indicate the distribution of temperature or some other quantity. The colours do not relate to the actual object.

False Cross Four stars in the far south of the sky: Kappa and Delta Velorum and Iota and Epsilon Carinæ, which form a pattern not unlike that of the *Southern Cross, though it is larger, more symmetrical, and contains no star as bright as the first magnitude. It has often been mistaken for the *Southern Cross. Three of its stars are white; the fourth (Epsilon Carinæ) is orange.

Fauth, Philipp (1867-1944) German amateur astronomer, who drew a large if rather inaccurate map of the Moon. Unfortunately he supported the absurd glaciation theory of the lunar surface, which harmed his scientific reputation.

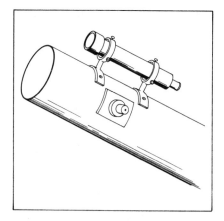

Finder telescope.

Faye's Comet A periodical comet with a period of 7.4 years, discovered by the French astronomer Hervé Faye in 1843 and seen at most returns since then.

Filaments *Prominences seen silhouetted against the Sun's disk, when they appear as dark threads.

Filar micrometer A device used for measuring very small distances as seen in a telescope. In its simplest form, it consists of two fine wires (often spider-threads), one of which is fixed, while the other may be moved by means of a screw. Filar micrometers are widely used for measuring the separations between the components of *double stars.

Finder A small telescope attached to a larger one. The finder has low magnification, but it has a wide field of view, which makes objects easy to locate; it is much more difficult to get an object into the relatively small field of the larger telescope. The procedure is to bring the object to the centre of the finder field; when this has been done, the object should also be visible in the main telescope, always provided that the finder is correctly lined up!

Fireball An exceptionally bright *meteor, having a *magnitude of −5 or brighter.

Micrometer.

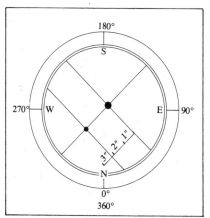

First Point of Aries The *vernal equinox. See *Ecliptic.

First Point of Libra The autumnal equinox. See *Ecliptic.

Fixed stars An old name for ordinary stars, to distinguish them from the 'wandering stars' or planets. The term is now obsolete.

Flammarion, Camille (1842-1925) French astronomer, noted for his studies of Mars and for his popular books and articles.

Flamsteed, John (1646-1719) English clergyman, who became the first Director of *Greenwich Observatory in 1675 and was subsequently created Astronomer Royal. His main contribution was in drawing up a star catalogue which was much the best of its time.

Flare stars Faint red dwarf stars which may brighten up noticeably over a period of a few minutes, fading back to their normal brightness within an hour or so. It is thought that this behaviour must be due to intense flare activity in the star's atmosphere. The brightest flare star is *UV Ceti.

Flares, Solar Brilliant outbreaks in the solar atmosphere, usually associated with active sunspot groups. A typical flare reaches its maximum brightness in a few minutes, and takes at least an hour to die away. Flares are composed of hydrogen, and send out electrified particles and large amounts of short-wave radiation, which reach the Earth and cause *magnetic storms, *aurorae, and interference with radio communications. Generally they are observed by using instruments based upon the principle of the spectroscope, though a few have been seen in integrated light.

Flash spectrum Just before the Moon completely covers the Sun at a total solar *eclipse, the Sun's *chromosphere is seen shining by itself, without the usual brilliant background. The dark lines in the spectrum then become bright, producing what is termed the flash spectrum. The same effect occurs just after the end of totality.

Fleming, Wilhemina (1857-1911) Scottish astronomer, who spent much of her life in America and joined the staff of the Harvard Observatory. She was closely involved with the famous *Draper Catalogue; she also discovered 10 novae and over 200 variable stars.

Flocculi Patches on the Sun's surface, observed by instruments based on the

Loop prominence on the Sun, 29 August 1970 (R. Lane).

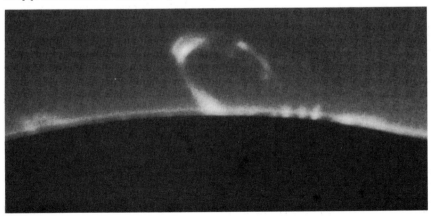

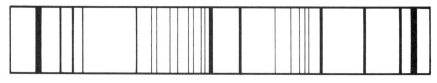

Absorption or Fraunhofer lines.

principle of the spectroscope. Bright flocculi are made up of calcium, dark flocculi of hydrogen.

Flying saucers Unidentified flying objects (UFOs) reported at intervals since 1947. They are due to natural phenomena such as clouds, ice crystals, the planet Venus, weather balloons and searchlight beams.

Focal length The distance between the centre of a mirror (or lens) and the *focus.

Focal ratio The ratio of the *focal length of a lens (or mirror) to the aperture; if f indicates the focal length and d the diameter of the lens or mirror, the focal ratio is equal to f/d.

Focus The point where the rays of light meet after being converged by a lens or mirror.

Fœcunditatis, Mare (The Sea of Fertility) One of the major lunar seas, easily visible with the naked eye. It leads off the Mare *Tranquillitatis.

Fomalhaut The star Alpha Piscis Australis; for data, see *Stars. Fomalhaut is the southernmost of the first-magnitude stars to be visible from Britain. In 1983 the infra-red satellite *IRAS discovered that it has an infra-red excess, due to cool material which is associated with it and which may possibly be planet-forming.

Forbidden lines Certain lines in the spectra of celestial bodies which do not usually appear under normal laboratory conditions, where materials are so much less rarefied.

Fraunhofer, Joseph von (1787-1826) German optician, who mapped the dark lines in the solar spectrum. His object-

glasses were much the best of their time, and he also made important contributions to theoretical optics.

Fraunhofer lines The dark *absorption lines in the spectrum of the *Sun.

Free fall The normal state of motion of an object in space under the gravitational influence of the pull of a central body; thus the Earth is in free fall round the Sun. An artificial satellite orbiting the Earth is in free fall, and a man inside it will have no sensation of 'weight'; he will be experiencing conditions of *zero gravity.

Free-free radiation The radiation emitted by a free electron as it is accelerated in the electrostatic field of an *ion.

Freyja Montes Mountains adjoining *Ishtar Terra on Venus.

Friedman, Alexander (1888-1925) Russian mathematician, who made very important contributions to relativity theory and to cosmology.

Frigoris, Mare (The Sea of Cold) An irregular sea in the northern hemisphere of the Moon. It leads off from the Sinus *Roris on one side and the Lacus *Somniorum on the other.

Fundamental stars Stars whose positions have been determined with the greatest possible accuracy, so that they can act as reference points in the measurements of the positions of other stars.

G

Gagarin, Yuri (1934-1968) The first space-man; in April 1961 he orbited the

Earth in *Vostok 1. He was killed in an aircraft crash.

Galactic co-ordinates These are based upon the galactic equator, which follows a line more or less represented by the *Milky Way and is inclined to the *celestial equator by 62°. The galactic longitude of a celestial body is the angular distance north or south of the galactic equator, measured along the great circle passing through the body and both galactic poles (the north galactic pole lies in Coma Berenices, the south pole in Sculptor). Galactic longitude is the angular distance, from 0° to 360°, measured eastward from the direction of the galactic centre—taken as being RA 17 hours 42.4 minutes, declination −28°55′ in the constellation of Sagittarius.

Galaxies Independent star-systems, often made up of many thousands of millions of stars together with interstellar material in the form of gas and 'dust'. About a thousand million of them are within the range of modern instruments, but only three are clearly visible with the naked eye; the southern *Magellanic Clouds, and the northern *Andromeda Spiral, M.31.

Galaxies are of various shapes. Many are spiral in form, and look like Catherine-wheels when observed (or, better, photographed) telescopically; others are elliptical or spherical, and there are also galaxies which are irregular in form. It used to be thought that the different shapes indicated different stages in evolution, but this idea is not now accepted.

The system of classification introduced by E. *Hubble is still in use, though more detailed systems have also been worked out. There are spirals of different degrees of 'tightness' (Sa, Sb, Sc); spirals in which the arms seem to come from the ends of a 'bar' through the main plane (barred spirals, SBa, SBb, SBc) and ellipticals, ranging from E0 (to all intents and purposes globular) to E7 (extremely flattened). *Seyfert galaxies have condensed nuclei and weak spiral arms; *Markarian galaxies are very powerful in the ultra-violet range. Our own *Galaxy appears to be a rather loose spiral of type Sb.

Apart from over two dozen relatively close systems making up the so-called *Local Group, all the galaxies appear to be receding from us, as is shown by the *red shifts in their spectra; the whole universe is expanding. The further away a galaxy is, the faster it is receding.

Many galaxies are highly active, and

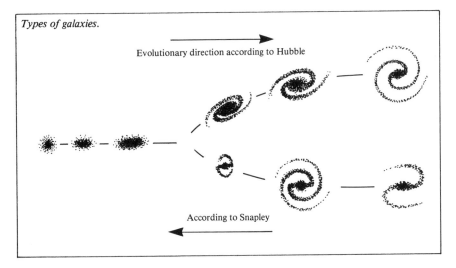

Types of galaxies.

Evolutionary direction according to Hubble

According to Snapley

M.16, a reflection nebula in Serpens.

Above *Conrad (left) and Kerwin on EVA to free a solar panel on Skylab.*

Below *The M.20 Nebula in Sagittarius.*

The North America Nebula near Deneb, so-called because its shape resembles that of the North American continent.

unspectacular objects as seen in ordinary telescopes. Even the Andromeda spiral appears as nothing more than a dim, hazy patch. Photographs taken with giant instruments are needed to bring out their true forms really well.

Galaxy, The The system of stars of which our Sun is a member. It contains about 100,000 million stars, arranged in a shape which has been likened to that of two fried eggs clapped together back to back! The Sun lies near the edge of a spiral arm. The diameter of the Galaxy is about 100,000 light-years, and the maximum breadth about 20,000 light-years. The Sun lies about 30,000 light-years from the centre of the system (the exact distance is still rather uncertain); we cannot see the galactic centre directly, because it lies beyond the star-clouds in Sagittarius, and there is too much obscuring material in the way. The whole Galaxy is rotating; the Sun takes about 225 million years (or one *cosmic year) to complete a full revolution.

Galilean Satellites The four large satellites of Jupiter: *Io, *Europa, *Ganymede and *Callisto.

Galilean telescope The first type of *refractor, made up of a single long-focus *object-glass and a negative *eyepiece. It gives an erect image.

Galilei, Galileo (1564-1642) Always known simply as Galileo. The first great

Group of galaxies in Pegasus (Mount Wilson and Palomar).

M.101, a comparatively nearby galaxy.

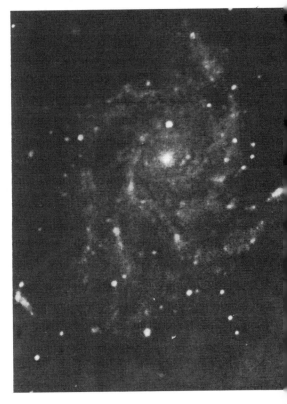

telescopic observer, and also the real founder of experimental mechanics. His championship of the *Copernican theory led to his trial and condemnation by the Inquisition in Rome, and the rest of his life was spent under strict supervision.

Galileo Space-craft An American vehicle, due to be launched in 1986. En route, it will take pictures of the minor planet *Amphitrite, but its main target is Jupiter. The Galileo probe will consist of two sections: an entry probe, which will plunge into Jupiter's clouds and (we hope) send back information before being destroyed, and an orbiter, which will be put into a closed path round Jupiter to survey both the planet and its system of satellites.

Galle, Johann (1812-1910) German astronomer. He and H. L. *D'Arrest were the first to identify *Neptune, in 1846, on the basis of calculations by U. J. J. *Le Verrier.

Gamma-ray astronomy The ultra-short gamma-rays are very difficult to study from ground level, because they cannot pass through the Earth's atmosphere, so that for most lines of investigation artificial satellites and space vehicles have to be used; it was only in 1969 that the first discrete cosmic gamma-ray source was identified. The most powerful sources are the *Vela pulsar and the *Crab Nebula. Gamma-rays come from the *Milky Way, but their origin is not yet fully understood. Only the very high-energy gamma-rays, with wavelengths a million million times shorter than that of light, can reach the ground; the main instrument for studying them is the gamma-ray telescope at Mount Hopkins in Arizona. Gamma-ray bursters have also been detected, but again we are not sure of their origin. It must be admitted that gamma-ray astronomy is still in its very earliest stage.

Gamow, George (1904-1968) Russian astrophysicist, who worked successively at Göttingen, Copenhagen, Cambridge and finally the United States. He made fundamental contributions to studies of stellar evolution, and he was also a brilliant biochemist. It is fair to say that his somewhat eccentric character led his contemporaries to be cautious about his results, but the value of his work is now recognized by all.

Gamma Ray Telescope at Mount Hopkins, Arizona, as I photographed it in 1982.

Ganymed Asteroid No 1036. It is an *Amor asteroid, discovered by W. *Baade in 1924; its period is 4.3 years. Its diameter is about 22 miles, so that it is the largest of the known Amor asteroids.

Ganymede The third and largest satellite of Jupiter; it is in fact the largest satellite in the Solar System. For data, see *Satellites. It is a very easy telescopic object, and there is firm evidence that very keen-sighted people can glimpse it with the naked eye.

Ganymede has an icy, cratered surface, as was shown by the *Voyager probes. There is probably a silicate core, overlaid by a mantle of water and ice. There are darkish regions on the surface, the largest of which, Galileo Regio, is nearly 2,500 miles in diameter; there are also strange grooves or furrows, together with low ridges. Ganymede has no detectable atmosphere, and has certainly been inert for a very long time, though there is some evidence of past crustal activity, and the surface may not be so ancient as that of *Callisto.

Garnet Star William *Herschel's nickname for the semi-regular or irregular variable Mu Cephei, which has an M2-type spectrum and is exceptionally red. Optical aid is needed to bring out the colour well; the magnitude range is from about 3.6 to 5, but the usual magnitude is around 4.5. Mu Cephei is over 1,500 light-years away, and must be at least 50,000 times as luminous as the Sun—much more powerful than *Betelgeux.

Gassendi A 55-mile lunar crater on the northern edge of the Mare *Humorum; it has a low central peak and a system of *rills on its floor. The north wall has been partly destroyed by the Mare lava. It is named in honour of the 17th century French astronomer Pierre Gassendi, who in 1631 was the first to observe a *transit of Mercury.

Gauss The standard unit of measurement for a magnetic field. The Earth's magnetic field, at the surface, ranges between 0.3 and 0.6 gauss.

Gauss, Karl (1777-1855) German mathematician—one of the greatest in scientific history. His calculations led to the recovery of the minor planet *Ceres a year after Piazzi had found it in 1801.

Gegenschein A very faint glow in the sky, exactly opposite to the Sun. It is excessively difficult to observe, and from Britain it is seldom seen; in size, it may cover an area equal to that of the Square of Pegasus. Like the *Zodiacal Light, it is due to sunlight illuminating the thinly-spread interplanetary material. It is sometimes called by its English name of the Counterglow.

Geiger counter An instrument used to detect charged particles and high-energy radiation. Most artificial satellites and space-vehicles carry Geiger counters.

Geminid Meteors A rich annual shower, with its maximum about 14 December. There is no known parent comet, though the orbit of the stream is similar to that of the asteroid *Phæthon. The *ZHR of the Geminids may be as high as 58.

Gemini programme Series of two-man space flights in the 1960s.

Geocentric theory The old theory that the Earth lay in the centre of the Solar System, with the Sun, Moon and planets moving round it.

Geodesy The study of the figure, mass, dimensions and other characteristics of the *Earth.

Georgian Planet An obsolete name for *Uranus.

Giacobini-Zinner Comet A periodical comet, originally discovered in 1900; the period is 6.5 years, and at its best the comet is an easy telescopic object. It is the parent comet of the *Giacobinid or Draconid meteors. In 1985 the American ICE (International Comet Explorer) probe passed through its tail.

Giacobinids (or Draconids) A meteor shower, associated with Comet

*Giacobini-Zinner; there were brilliant displays in 1933 and 1946, but in most years the shower is non-existent.

Giant Star A star to the upper right of the *Main Sequence on the *Hertzsprung-Russell Diagram.

Gibbous phase The phase of the Moon when between half and full. Mercury and Venus, of course, show lunar-type phases, and Mars may also appear decidedly gibbous at times.

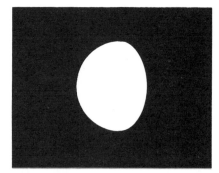

Gibbous phase.

Gill, Sir David (1843-1914) Scottish astronomer, who became HM Astronomer at the Cape in 1879. He determined the solar *parallax, and was a pioneer in the photographic mapping of the sky—a project begun after 1882, when he recorded many stars on a photograph he had taken of the bright comet of that year, and realized the immense potentiality of photographic surveys.

Gioja The north polar crater of the Moon. It is 22 miles in diameter.

Giotto Probe The European Space Agency probe to *Halley's Comet. It was launched on 2 July 1985 from the Kourou rocket base, in French Guiana, by an *Ariane rocket. The rendezvous date was 13-14 March 1986.

Glenn, John (1921-) The first American astronaut to orbit the Earth in 1962. This was his only flight. He became a Senator in 1974, and in 1984 offered himself as a Presidential candidate, though without success.

Globular clusters See *Clusters, Stellar.

Globules See *Bok globules.

Gnomon The 'pointer' of a *sundial; its function is to cast its shadow on the dial in order to show the time. The gnomon must always point toward the *celestial pole.

Goldstone The Californian tracking station operated by the *Jet Propulsion Laboratory.

Goodricke, John (1764-1786) English astronomer, born in Holland. He discovered the variability of *Delta Cephei, and explained the cause of the variation of *Algol. Goodricke was a deaf-mute; despite this, he would certainly have had a brilliant scientific career if he had lived.

Gould's Belt A belt of bright stars, inclined to the *Milky Way by about 20°. It was first pointed out by Sir John *Herschel in the 19th century, and was studied by the American astronomer B. Gould; it includes most of the bright stars in Orion, Scorpius, Carina and Centaurus as well as some bright stars in other constellations. It seems to be due to a slight tilting of the spiral arm of the Galaxy in the neighbourhood of the Sun, which actually lies just clear of the arm. The effect is most noticeable for the young stars of spectral type B.

Granules, Solar The Sun's bright surface or *photosphere is not smooth; when seen in detail, a granular structure appears, each granule being several hundreds of miles in diameter. The granules are short-lived, each lasting for only a few minutes, and are in constant motion. The best photographs of them have been taken by instruments carried in balloons and in artificial satellites.

Grating See *Diffraction grating.

Gravitation The force of attraction which exists between all particles of matter in the universe. Newton's Law, published in his great book in 1687, states that F (the attractive force between two bodies) is proportional to the product of their masses (m_1 and m_2) and inversely proportional to the square of the distance d between them, so that if G is the gravitational constant, $F = Gm_1m_2/d^2$.

Gravitational waves Wavelike effects produced by the disturbance of a massive body. They are better described as ripples in the curvature of spacetime. Many attempts have been made to record them, notably by J. Weber, but as yet there is no certain proof of their existence.

Grazing occultation An *occultation of a star by the Moon, when the star grazes the

The Giotto probe being prepared for its trip to Halley's Comet.

Moon's limb and is briefly hidden by a succession of mountain peaks.

Great circle A circle on the surface of a sphere (such as the *Earth) whose plane passes through the centre of the sphere. Thus a great circle will divide the sphere into two equal parts. The horizon, the celestial equator and the ecliptic are all great circles on the *celestial sphere.

Greek alphabet See *Stars.

Green Bank A major radio astronomy observatory in West Virginia.

Green Flash (or Green Ray) When the Sun is on the point of setting, its last visible segment may flash brilliant green for a moment. This is due to effects of the

Left *The 210-ft radio antenna at the Goldstone, California, complex of tracking stations* (JPL).

Below *An 85-ft antenna of the Goldstone station of the Deep Space Net in California's Mojave Desert* (JPL).

Earth's atmosphere, and is best seen over a sea horizon. Venus has also been known to show a Green Flash when setting.

Green Ray See *Green Flash.

Greenhouse effect The raising of the surface temperature of a body by the trapping of infra-red radiation—as with Venus, for example, where the greenhouse effect is caused by the large quantities of carbon dioxide in the planet's atmosphere.

Greenwich Mean Time (GMT) The local time at Greenwich, reckoned according to the *mean sun. It is used as the standard throughout the world.

Greenwich Meridian The line of longitude which passes through Greenwich Observatory (or, to be more precise, through the Airy transit circle there). It is still taken as longitude 0°, though the working instruments at Greenwich have now been moved to Herstmonceux in Sussex.

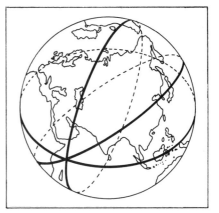

Great circles on the Earth's surface.

Greenwich Observatory The most famous British observatory, founded in 1675 by order of King Charles II—mainly so that an improved star catalogue could be drawn up for use of marine navigators. Until the retirement of Sir Richard Woolley, in 1971, the Director of the Observatory was also Astronomer Royal.

The old Royal Observatory in Greenwich Park.

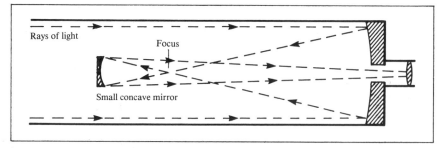

Principle of the Gregorian reflector.

Deteriorating conditions, due to the spread of London, led to the transfer of the equipment to *Herstmonceux in Sussex, and the original Greenwich Observatory is now a museum.

Gregorian Calendar The *calendar now in use.

Gregorian reflector A type of reflecting telescope in which the light from the object under study strikes the main mirror, and is reflected back up the tube on to a small concave mirror placed outside the focus of the main mirror. The light then comes back through a hole in the main mirror, as with the *Cassegrain, and reaches the eyepiece. Telescopes of this kind were first described by James Gregory before *Newton built his first reflector; they were popular for a while, but are not easy to make or adjust, and few of them are now in use. A Gregorian, unlike a Cassegrain or a Newtonian reflector, gives an erect image.

Grigg-Skjellerup Comet A periodical comet discovered by the New Zealand amateur John Grigg in 1902. It was recovered by Skjellerup in 1922, and since then has been seen at every return; it can develop a short tail. The period is 5.1 years, shorter than that of any other known comet apart from *Encke's. Since 1964, close encounters have occured between the comet and the Earth leading to the occurence of a new shower of *meteors.

Grimaldi A large lunar walled plain, 120 miles in diameter, named after the 17th

century Italian astronomer F. Grimaldi. The formation lies near the Moon's western limb, and has a floor so dark that it can be instantly recognized whenever it is in sunlight.

Gruithuisen, Franz von Paula (1774-1852) German astronomer, who carried out valuable observations of the Moon and planets; he was the originator of the impact theory of lunar cratering. Unfortunately, his vivid imagination tended to expose him to ridicule; for instance he announced the discovery of a 'lunar city' with 'dark gigantic ramparts', and believed the *Ashen Light of Venus to be due to artificial illuminations there.

Gum Nebula The *Vela supernova remnant, named in honour of the Australian astronomer Colin Gum. Its near side is little over 300 light-years away. The rate of expansion of the gas-cloud indicates that the supernova could have appeared about the year BC 9000, and would have become as brilliant as the half-moon.

H

H.I and H.II regions Clouds of hydrogen in the *Galaxy (and in other galaxies). In H.I regions the hydrogen is neutral (that is to say, each atom is complete) and the clouds cannot be seen visually, though they can be detected at radio wavelengths. In H.II regions the hydrogen is *ionized (that is to say, each atom is incomplete), and the presence of hot stars makes the cloud shine as a *nebula. A really hot star may have an effect stretching out to as much as 500 light-years from it.

HR Delphini A nova discovered in 1967 by the English amateur G. E. D. Alcock. It reached magnitude 3.6, and had an unusually long maximum. In 1985 it was still above its pre-outburst magnitude of about 12.

HR Diagram See *Hertzsprung-Russell Diagram.

Hadley Rill A lunar rill in the foothills of the *Apennines. Apollo 15 landed in this region, and Astronauts Scott and Irwin drove right up to the edge of the rill in their Lunar Rover.

Haemus Mountains Lunar mountain

range, forming the southern border of the Mare *Serenitatis.

Halation ring A ring sometimes seen round a star image on a photograph. It is, of course, purely a photographic effect.

Hale, George Ellery (1868-1938) American astronomer, noted for his solar work; he invented the *spectrohelio-graph. he was also mainly responsible for the setting-up of great telescopes, the last of which was the 200-in at Palomar—known, very appropriately, as the Hale Reflector. It was completed in 1948, and was for many years the largest telescope in the world; it is still surpassed in size only by the Russian 236-in.

Hall, Asaph (1829-1907) American

Orbit of Halley's Comet.

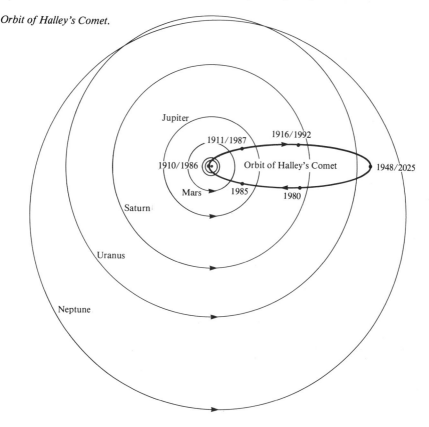

astronomer who carried out much important work, but is best remembered for his discovery of Phobos and Deimos, the two satellites of Mars, in 1877.

Halley, Edmond (1656-1742) The second Astronomer Royal, succeeding *Flamsteed. Halley went on an expedition to St Helena to study the southern stars, and undertook important researches in all fields of astronomy as well as in other sciences; he was also entirely responsible for the publication of *Newton's great book the *Principia*, in 1687. He predicted the return of the periodical comet which now bears his name.

Halley's Comet The only bright comet whose return can be predicted. It has a period of 76 years, and last came to perihelion on 9 February 1986.

Records of Halley's Comet go back for many centuries. It was certainly recorded by the Chinese in BC 1059, and there is a strong chance that it is the comet referred to in a report from BC 2467. Every return since that of BC 240 has been observed. Sometimes the comet has been brilliant—notably in 837, when the head was as bright as Venus and the tail stretched for more than 90° across the sky. There was a return in 1066, just before the Norman invasion of England; the comet is shown in the famous Bayeux Tapestry, with the Saxon courtiers looking on aghast and King Harold toppling on his throne! The Italian painter Giotto di Bondone saw it in 1301, and used it as a model for the Star of Bethlehem in his picture *The Adoration of the Magi* (which is why the European probe sent to the comet in 1985-6 was named '*Giotto'). In 1456 there is a story, which may or may not be true, that Pope Calixtus III condemned it as an agent of the Devil! Edmond *Halley observed it in 1682, and realized that it was identical with the comets of 1607 and 1531, so that he could predict the next return with confidence (though he did not live to see it). Unfortunately the 1986 return was the least favourable for many centuries, and the next, that of 2061, will be no better. Halley's Comet is associated with two meteor showers, the *Eta Aquarids and the *Orionids.

Halo The spherical-shaped star cloud round the main *Galaxy. It is best referred to as the *galactic halo*, to distinguish it from a meteorological halo—a luminous ring observed round the Sun or Moon, due to ice crystals in the Earth's upper atmosphere.

Hamal The star Alpha Arietis. For data, see *Stars. Its orange colour is very evident with binoculars.

Harriot, Thomas (1560-1621) English scholar, at one time tutor to Sir Walter Raleigh, who drew the first telescopic map of the Moon, some months before *Galileo.

Harvard classification The original classification of stars into spectral types, denoted by letters of the alphabet. In its modified form it is still in use (see *Stars).

Harvest Moon In the northern hemisphere, the full moon closest to the autumnal *equinox is termed Harvest Moon; the autumnal equinox falls on 22 September or thereabouts.

Usually the Moon rises more than half an hour later from one night to the next; this time-lapse is known as the *retardation. It may sometimes amount to over an hour, so that if, say, the Moon rises at 22:00 GMT on one evening it may not rise until 23:00 GMT on the next. Near the autumnal equinox, however, the retardation is no more than 15 minutes—though it is incorrect to say, as many books do, that the Moon then rises at almost the same time on several evenings in succession. It is also wrong to suppose that the Harvest Moon looks any larger than any other full moon; it does not.

Hathor An *Aten-type asteroid, with a period of 0.76 year. It is only a mile or two in diameter, and is always very faint. It was the third asteroid of its type to be discovered, following *Aten itself and *Ra-Shalom.

Haute-Provence Observatory Major French observatory, equipped with a 74-in reflector.

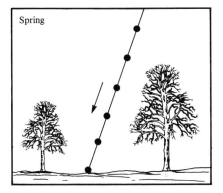

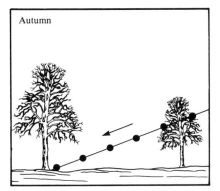

Harvest Moon. In the spring (northern hemisphere) the ecliptic makes a low angle with the horizon. The black circles indicate the shift of the Moon over 24 hours. In the autumn, the ecliptic makes a sharper angle with the horizon, so that the daily retardation is much less.

Hawking, Stephen (1942-) Brilliant English mathematician and astronomer, who has made fundamental contributions to cosmology and allied subjects despite the handicap of a crippling physical disease.

Hayashi Track A downward track on the *Hertzsprung-Russell Diagram which is followed by very young stars still contracting toward the *Main Sequence. It was named after the Japanese astronomer Chushiro Hayashi (1920-).

Hector A *Trojan asteroid. Its diameter is probably over 100 miles, but it may be irregular in shape, or even double.

Heliacal rising The rising of a celestial body at the same time as sunrise. The more common use of the term is, however, the date when the body first becomes visible in the dawn sky. For timekeeping purposes, the ancient Egyptians paid great attention to the annual heliacal rising of *Sirius.

Heliocentric theory The theory according to which the Sun lies at the centre of the Solar System, as proposed by *Copernicus—and, very much earlier, by the Greek philosopher *Aristarchus. The word comes from the Greek *helios* (= Sun).

Heliometer A refracting telescope in which the object-glass is cut in half, so that one half may be made to slide past the other—giving a double image of the object under study. It is used to measure very small distances as seen through the telescope eyepiece, so that its purpose is the same as that of a *micrometer even though the principle is quite different.

Heliostat A *cœlostat mirror, rotatable so that it can reflect light from the target object in a fixed direction.

Helium The second lightest *element (after *hydrogen). Apart from hydrogen it is much the most plentiful element in the universe. It was discovered in the solar spectrum, in 1868, well before it was first identified on Earth.

Helium flash The sudden start of helium burning (by the *triple-alpha process) in the *degenerate core of a star in the later stages of its evolution.

Helium star An old, obsolete name for a B-type star.

Helix Nebula NGC 7293, a bright *planetary nebula in Aquarius. Its distance is about 450 light-years. It is visible in binoculars, and is therefore much brighter than the more famous M.57, the *Ring Nebula in Lyra.

Hellas The largest and deepest basin on Mars; it measures about 1,370 miles by 1,120 miles. It was once believed to be a snow-covered plateau, and it can sometimes appear as brilliant as the polar cap. Its floor is relatively lacking in detail.

Henbury Craters *Meteorite craters in Australia. There are thirteen in all, the largest of which is over 600 ft in diameter.

Henderson, Thomas (1798-1844) Scottish astronomer. While Director of the Cape Observatory, in the 1830s, he measured the *parallax of *Alpha Centauri, but did not work out the star's distance until after F. W. *Bessel had announced the distance of the star 61 Cygni. In 1834 Henderson became the first Astronomer Royal for Scotland.

Heraclides (BC 388-315) Greek philosopher. His writings are lost, but it seems that he believed the Earth to rotate on its axis in a period of 24 hours.

Herbig-Haro objects Small, bright condensations of dust and gas, some of which are to be found in the *Orion Nebula. They show detectable changes in form over periods of some years, and may be very young stars condensing toward the *Main Sequence. They are often found in the neighbourhood of *T Tauri variables.

Hercules X-1 An X-ray binary; the visible star is known as HZ Herculis, while the secondary is probably a *neutron star.

Principle of the Herschelian reflector.

Hercules X-1 was the first X-ray binary to be discovered.

Hermes A very small *Apollo asteroid. In 1937 it passed within 485,000 miles of the Earth—the closest approach of any natural object apart from the Moon. Its period was given as 2.1 years, but it has not been seen again, and its recovery will be largely a matter of luck. It has not been given a permanent asteroid number; its provisional designation was 1937 UB.

Herschel A crater on Saturn's satellite *Mimas. It is 80 miles in diameter; the diameter of Mimas itself is only 243 miles, so that Herschel dominates the entire scene.

Herschel, Caroline (1750-1848) Sister and colleague of *William Herschel. She was herself an able observer who discovered six comets.

Herschel, Sir John (1792-1871) Son of *William Herschel. He too was a great observer; from 1833 to 1838 he was at the Cape of Good Hope, making the first really systematic telescopic studies of the far-southern stars. He was also the last to see *Halley's Comet at its 1835 return.

Herschel, Sir William (1738-1822) Probably the greatest of all observers. Born in Hanover, he came to England as a young man and became a professional organist at Bath; in 1781 he discovered the planet *Uranus, and thenceforth devoted his life to astronomy, becoming official astronomer to King George III (not Astronomer Royal; that post was held by

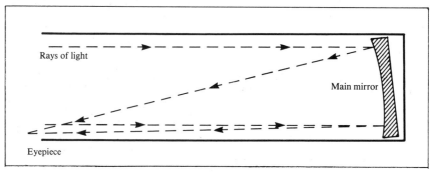

*Maskelyne). Herschel discovered thousands of new double stars, clusters and nebulæ; he established the nature of *binary systems; his main work was in connection with the distribution of the stars, and he was the first to provide a reasonably accurate picture of the shape of the *Galaxy. He even speculated on the possibility that the so-called 'starry nebulæ' were external galaxies in their own right. He was the best telescope-maker of his time; his largest reflector had a focal length of 40 ft, and a 49-in mirror.

Herschel-Rigollet Comet A periodical comet discovered in 1788 by *Caroline Herschel. It has a period of 156 years, and was recovered in 1939 by R. Rigollet. With the exception of Comet Grigg-Mellish, it has the longest period of any comet which has been observed at more than one return. It can attain the 7th magnitude, and develops a short tail when near perihelion.

Herschelian reflector A type of reflecting telescope developed by *William Herschel in the late 18th century. The main mirror is tilted, and the light is brought to focus at the side of the tube, thus removing the need for any secondary mirror. However, there are numerous disadvantages to this system, and Herschelian or 'front-view' reflectors are now seldom used.

Herstmonceux The present site of the Royal *Greenwich Observatory, near Hailsham in Sussex. The old Castle is used as offices. There are various working telescopes, though the largest, the *Isaac Newton reflector, has now been moved to a better site at *La Palma in the Canary Islands.

Hertzsprung, Ejnar (1873-1967) Danish astrophysicist; one-time Director of the Leiden Observatory in Holland. In 1905 he discovered the giant and dwarf divisions of the stars.

Hertzsprung-Russell Diagram (Often known simply as the HR Diagram.) A diagram in which the stars are plotted according to their spectral types and their

The Isaac Newton Telescope and interior of the dome at Herstmonceux. The telescope is now at La Palma.

*absolute magnitudes. As described under the heading *Star, the stars are divided into various classes: O, B and A (white), F and G (yellow), K (orange) and M, R, N and S (red).

When the diagram was compiled, by E. *Hertzsprung and H. N. *Russell before the First World War, a definite pattern was seen to emerge. Most of the stars lie on a well-defined belt known as the *Main Sequence, but there are also very luminous stars of types M and K making up a *giant branch*, while near the lower left-hand portion of the diagram there are the hot but very small *White Dwarfs. The Sun is a typical G-type Main Sequence star.

81

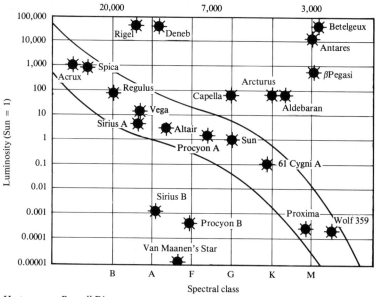

Surface temperature   °K

Hertzsprung-Russell Diagram.

HR Diagrams have been of the utmost value in theoretical astrophysics. It was once thought that a star began as a giant, shrank and heated up until joining the Main Sequence at the upper left (type O or B) and then passed down the Main Sequence, ending its career as a red dwarf or type M; but it is now known that this whole approach is wrong. For much of its career a star remains at much the same point on the Main Sequence, after which it moves into the giant branch before collapsing into a White Dwarf or a *neutron star. Instead of being youthful, red supergiants such as *Betelgeux are well advanced in their evolution.

Hesperus The ancient name for *Venus as an evening object.

Hevelius, Johann (Hewelcke) (1611-1687) Danzig astronomer, who drew up a notable star catalogue and compiled a map of the Moon. He had his private observatory in Danzig (the town now known as Gdańsk, in Poland).

Hexahedrite A form of iron *meteorite, containing between 4 and 6 per cent of nickel and consisting mainly of the cubic mineral kamacite.

Hidalgo Asteroid No 944, discovered by *Baade in 1920. It is no more than about 10 miles in diameter, and has a very eccentric orbit; its distance from the Sun ranges between 186 million miles and 894 million miles, so that at *aphelion its distance is greater than the mean distance of Saturn. The period is 14 years—with the exception of *Chiron, the longest of any known asteroid. There have been suggestions that Hidalgo may be a cometary nucleus, but no trace of coma has ever been observed.

High-velocity stars Old *Population II stars in the galactic *halo. They move in elliptical orbits which carry them well away from the galactic plane. They are not genuinely moving faster than other stars, but they seem to have higher velocities because of their greater speed relative to the Sun.

Himalia The sixth satellite of Jupiter; for data, see *Satellites. Himalia's diameter is about 115 miles, so that it is much the largest of the Jovian satellites apart from the four *Galileans. Its mean opposition magnitude is 14.8, so that it is not an easy telescopic object.

Hipparchus (BC 190-120) Greek astronomer, who drew up a star catalogue upon which *Ptolemy's was based. Hipparchus also discovered *precession, and made important advances in mathematics. Little is known of his life, and his original catalogue has not come down to us.

Hipparchus A low-walled, 90 mile crater not far from the apparent centre of the lunar disk as seen from Earth. It is adjoined to the south by the smaller but better-formed formation Albategnius.

Hirayama families Groups of *minor planets whose members have similar orbits. More than forty such families are known. The name honours the Japanese astronomer K. Hirayama, who first drew attention to them in 1928.

The Horse's Head Nebula.

Hoba West Meteorite The largest meteorite known; its weight is at least 60 tons. It is still lying where it fell, in prehistoric times, at Hoba West, near Grootfontein in Southern Africa. It has produced no crater.

Homestake Mine A working gold-mine in South Dakota. In it, at a depth of a mile, is a solar observatory—consisting of a large tank of cleaning fluid which is used to detect *neutrinos from the Sun. At present, Dr Ray Davies and his colleagues at Homestake find that there are fewer solar neutrinos than theory predicts. The detector has to be beneath a mile of rock, as otherwise it would be impossible to separate the effects of neutrinos from those of *cosmic rays, which are less penetrating than neutrinos and so cannot reach the detector.

Hooker reflector The *Mount Wilson 100-in reflector, financed by the millionaire J. Hooker at the instigation of G. E. *Hale. It was responsible for fundamental advances in our knowledge, and between its completion, in 1917, and the opening of the *Palomar reflector in 1948 it was not only the largest telescope in the

world, but was in a class of its own. Sadly, it was taken out of active commission in 1985.

Horizon The great circle on the *celestial sphere which is everywhere 90° from the observer's overhead point or *zenith. The actual horizon will not usually be quite the same as the apparent horizon, partly because of irregularities in ground level and partly because of the height of the observer himself.

Horrocks, Jeremiah (1619-1641) English curate and amateur astronomer, who in 1639 made the first observation of a *transit of Venus. His early death cut short a career of exceptional promise.

Horse's Head Nebula A dark *nebula in Orion, near the star Alnitak or Zeta Orionis. The name comes from its rather

The Mare Humboldtianum photographed by Mariner 10 on the probe's way to Venus.

obvious resemblance in shape to that of a knight in chess! It is by no means easy to see telescopically; photographs are needed to show it properly.

Hour angle The time which has passed since a celestial object crossed the *meridian. (The meridian is the great circle on the *celestial sphere which passes through the *zenith and both celestial poles, so that it cuts the horizon at the north and south points.) If the object has not yet crossed the meridian and is therefore in the eastern part of the sky, its hour angle is negative. To find the hour angle of a body, simply subtract its *right ascension from the local *sidereal time.

Hour circle A great circle on the *celestial sphere passing through both the celestial poles. The zero hour circle coincides with the observer's meridian.

Hoyle, Sir Fred (1915-) Great British astronomer, who has made many contributions to astrophysics and cosmology. He was one of the supporters of the

*steady-state theory, and is of the opinion that *quasars are much less remote and less powerful than is usually believed.

Hubble, Edwin (1889-1953) American astronomer, who worked extensively with the Mount Wilson 100-in *Hooker reflector. He made many notable contributions, but is best remembered for his work in connection with the distances of the *galaxies. In 1923 he discovered *Cepheids in some of the external systems, and was able to show that they lie far beyond our Galaxy. This was possibly the most important astronomical discovery of the present century.

Hubble's constant A constant relating to the recessional velocities of galaxies. The presently-accepted value is 55 kilometres per second per *megaparsec, but this may be considerably in error; estimates range between 50 km/sec/mpc and as much as 100 km/sec/mpc.

Hubble time The time which has elapsed since the origin of the universe between 15,000 and 20,000 million years ago.

Huggins, Sir William (1824-1910) English amateur astronomer; one of the great pioneers of stellar spectroscopy. He had his private observatory at Tulse Hill, in outer London.

Humason, Milton (1891-1972) American astronomer, who worked closely with *Hubble at Mount Wilson, and played a very major rôle in the investigations of the spectra of galaxies and the expansion of the universe. His first post at Mount Wilson was that of a mule driver; he attained his international reputation without any formal training.

Humboldtianum, Mare (Humboldt's Sea) A lunar sea near the Moon's north-eastern limb.

Humorum, Mare (The Sea of Humours) A well-defined lunar sea, leading off the Mare *Nubium. The large crater *Gassendi lies at its border, and two well-known bays, Doppelmayer and Hippalus, lead off it. There are no large craters on its floor.

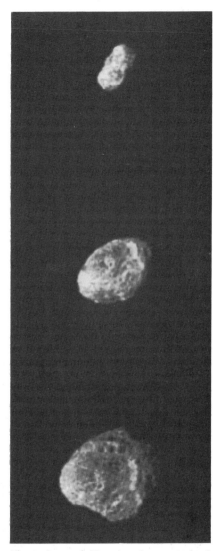

Three views of Hyperion (see next page) obtained as Voyager 2 flew past Saturn in 1981. The satellite is shaped roughly like a hamburger! (JPL).

Hunter's Moon The full moon following *Harvest Moon. Since Harvest Moon usually falls during the last part of September, Hunter's Moon occurs in October.

Huygens, Christiaan (1629-1695) Dutch scientist; one of the ablest observers of his time. He was the first to recognize the nature of Saturn's rings, and he discovered *Titan. Huygens also improved telescope designs, and made the first successful pendulum clock.

Hyades The open star *cluster near *Aldebaran. (Aldebaran is not a member of the cluster, but lies in the foreground.) The Hyades are easily visible with the naked eye as a kind of V pattern. The Hyades make up a *moving cluster, about 500 million years old; there are about 100 stars within a radius of eight light-years from the centre. At its distance of 140 light-years, this is the nearest of all star-clusters.

Hydroxyl radical The chemical *radical OH. It was the first molecule to be identified in interstellar space (in 1963).

Hygeia Asteroid No 10. Its diameter is 280 miles, and it is therefore larger than any other asteroid apart from *Ceres, *Pallas and *Vesta (excluding the exceptional *Chiron). Hygeia is classed as a 'peculiar carbon' asteroid. Its period is 5.6 years, and the mean opposition magnitude is 10.2.

Hyperion The seventh satellite of Saturn, moving between the orbits of *Titan and *Iapetus. It is irregular in shape, with a longest diameter of 249 miles and a shortest diameter of 149 miles. Its longer axis does not point directly at Saturn, as dynamically it ought to do. Hyperion is less reflective than the other icy satellites, and there may be a 'dirty' layer covering part of the surface. Several craters and one long ridge were shown by the *Voyager probes.

I

Iapetus The eighth satellite of Saturn; for data see *Satellites. It was the second satellite to be discovered (by Cassini, in 1671). Cassini noted that Iapetus is much brighter when west of Saturn than when to the east, presumably indicating a surface with unequal *albedo; the revolution period is 79 days, and the rotation is *captured. The *Voyager space-craft confirmed this. The leading hemisphere has an albedo of no more than 0.05, while the trailing hemisphere is conventionally bright, with an *albedo of about 0.5. Since the overall density of Iapetus is not much greater than that of water, we are clearly dealing with an icy globe stained in part by a dark deposit of unknown composition. Craters exist, though the Voyagers approached Iapetus less closely than with the other major satellites, so that the available maps are less detailed.

Icarus Asteroid No 1566, discovered by *Baade from Palomar in 1949. It is no more than a mile in diameter. Until the discovery of *Phæthon, in 1983, Icarus was the only known asteroid whose orbit carried it inside that of Mercury; the period is 1.12 years, and the minimum distance from the Sun is only about 17 million miles.

Ikeya-Seki Comet A bright comet seen in 1965. It was briefly spectacular as seen from some parts of the world, though not from Britain. The period has been estimated as about 880 years.

Image intensifier An electronic device used to enhance a dim optical image. Used with large telescopes it has proved to be remarkably effective.

Imbrium, Mare The largest of the regular lunar seas, easily visible with the naked eye. It is bounded in part by the *Apennines and the *Alps; it leads into the *Oceanus Procellarum. There are not many large craters on it, the principal formations being those of the Archimedes group.

Immersion The entry of a celestial object into *occultation or *eclipse. When a star is occulted by the Moon, immersion is practically instantaneous, because the Moon has no atmosphere, and a star is

Iapetus, Saturn's outermost large satellite, seen at a distance of 1.1 million km by Voyager 2 in 1981 (JPL).

practically a point source (the same is true of reappearance, or emersion).

Inferior conjunction When Mercury and Venus lie almost between the Earth and the Sun—that is to say, their *right ascension is the same as that of the Sun—they are said to be at inferior conjunction; they are of course 'new', and so cannot be seen unless the lining-up is exact enough to produce a *transit. Only bodies which are closer to the Sun than we are can reach inferior conjunction—that is to say Mer-

cury, Venus, some comets, and *Apollo and *Aten type asteroids.

Inflationary Epoch A short period in the very early history of the universe immediately following the *Big Bang, when the scale of the universe increased with unprecedented rapidity.

Infra-red astronomy Radiation with wavelength longer than that of red light (about 7,500 *Å) cannot be seen with the eye. This infra-red region extends up to the short-wave end of the radio part of the *electromagnetic spectrum (*micro-wave region).

Infra-red astronomy has become of immense importance in recent years.

There are many thousands of discrete infra-red sources in and beyond the Solar System; there are also wispy clouds of 'dust' scattered widely across the sky, known as infra-red cirrus, possibly consisting of graphite particles which have been formed in the outer atmospheres of stars and heated by stellar radiation. Infra-red radiation can be recorded from the region of the centre of the *Galaxy, and there are infra-red sources in nebulæ, such as the *Becklin-Neugebauer Object in the Orion Nebula. Some *galaxies which are very dim visually are very powerful at infra-red wavelengths. Much has been learned from infra-red satellites, notably *IRAS.

Infra-red studies are difficult from the surface of the Earth, because the wavelengths are absorbed by water-vapour in the atmosphere. Therefore the main infra-red telescopes are at high altitude. The largest is *UKIRT, on Mauna Kea in Hawaii, at an altitude of about 14,000 ft.

Innes, Robert Thorburn Ayton (1861-1933) Scottish astronomer, who worked first in Australia and then in South Africa. He was concerned largely with double stars, discovering 1,500 new pairs. He also identified *Proxima Centauri, the nearest star beyond the Sun.

Interferometer A stellar interferometer is an instrument used for measuring the diameters of stars; it depends upon the principle of light-interference. Radio interferometers are also used; see *Radio Astronomy.

Intergalactic matter Material spread between the galaxies. Evidently there is much more of it than used to be thought.

International Geophysical Year (IGY) The period between July 1957 and the end of December 1958, when scientists of over fifty nations co-operated in a programme of observations to learn more about all aspects of the Earth. The data obtained during the IGY took many years to analyze, and a tremendous amount of new knowledge was obtained.

Interstellar absorption The absorption of light by *interstellar matter. Light is not only dimmed, but is also reddened, so that distant stars in some directions look 'redder' than they really are. The main absorption is near the plane of the *Milky Way, where galaxies are concealed (see *Zone of Avoidance).

Interstellar matter Rarefied gas and 'dust' spread between the stars. In recent years, many types of molecules have been detected in interstellar clouds. The mean density of interstellar matter is very low indeed.

Invariable plane The plane through the centre of mass of the Solar System. It is inclined to the Sun's equator by about 7°.

Inverse Compton Effect See *Compton Effect.

Io The innermost of the *Galilean satellites of Jupiter. For data, see *Satellites.

Io is slightly larger than our Moon, with a diameter of 2,258 miles; its density, 3.55 times that of water, is greater than those of the other Galilean satellites. The *Voyager space-craft have shown that it is a remarkable world. Its surface is sulphur-coloured and bright red; there are 'hot spots', and violently active sulphur volcanoes, such as *Loki and *Pele. According to one theory, there is a crust of sulphur and sulphur dioxide, solid only in its outermost part, overlying a sulphur 'sea'. The interior is constantly flexed by Jupiter's gravitational pull, and to a certain extent by *Europa. There is an excessively tenuous atmosphere. Io has a marked effect upon the radio emissions from Jupiter, and is connected to Jupiter by a powerful electric flux-tube. It moves within the radiation zones round Jupiter, and must be a candidate for the title of the most lethal world in the Solar System!

Ion An *atom which has lost one or more of its *electrons, and so has a positive electrical charge, since in a complete atom the positive charge of the nucleus is balanced out by the combined negative charge of the planetary electrons. The process of producing an ion is known as *ionization*.

An atom which has gained an extra electron is *positively ionized*.

Ion tail The gaseous tail of a comet. Unlike the dust tail, it is usually almost straight.

Ionosphere The region of the Earth's atmosphere above the *stratosphere. It extends between about 40 and about 500 miles above the ground, and contains the layers which reflect some radio waves back to Earth, this making long range wireless communication possible.

Iota Aquarids A minor meteor shower, with its maximum about 6 August. The usual *ZHR is about 6.

IQSY (International Years of the Quiet Sun) Periods when combined efforts are made to study the Sun near the minimum of its cycle of activity. Many nations take part, as in the *IGY.

Comet IRAS-Araki-Alcock, 1983.

IRAS (Infra-Red Astronomical Satellite) A space-craft launched in January 1983; it operated until late in the year. It mapped thousands of new infra-red sources, as well as providing information about *dust-rings in the Solar System and tracking down infra-red excesses with various stars, such as *Vega, which could indicate planetary material. It discovered several comets, and also found dust-tails to some known periodical comets. All in all, IRAS has been one of the most successful satellites to date.

Iridum, Sinus The Bay of Rainbows, a beautiful lunar bay leading out of the Mare *Imbrium. When the Sun is rising over it, the solar rays first catch the mountains bordering the bay to the west, producing the effect nicknamed the 'Jewelled Handle'. The old boundary between the Sinus Iridum and the Mare Imbrium has been virtually destroyed, though traces of it remain between the two high capes of Laplace and Heraclides.

The international Infra Red Astronomical Satellite (IRAS) being tested (JPL).

Irradiation The effect which makes brilliantly-lit or self-luminous bodies appear larger than they really are.

Isaac Newton telescope The largest telescope ever made in Britain. It had a 98 in mirror, and in 1967 was brought into use at the Royal Greenwich Observatory, Herstmonceux. It has now been given a new, 101-in mirror, and transferred to the new observatory on *La Palma.

Ishtar Terra The second largest upland region on Venus, with a diameter of 1,800 miles. It lies in the northern hemisphere of the planet; its western part, Lakshmi Planum, is a smooth plateau, while to the eastern end are the *Maxwell Mountains, the highest on Venus.

Island universes An obsolete name for *galaxies.

Isophote A line on a diagram joining points of equal intensity or density.

Isotropy Looking the same in all directions.

IUE (International Ultra-violet Explorer) A space-craft launched from Cape Canaveral in January 1978. It was very successful, and provided a tremendous amount of information about ultra-violet sources of all kinds. It was a joint venture by NASA, ESA and Britain. Results include the first ultra-violet spectrum of a *supernova, the first high-resolution ultra-violet spectrum of a star in an external galaxy, and observations of the galactic 'corona', a hot gaseous envelope extending out to 25,000 light-years and with a temperature of around 100,000°C.

Ivory Coast Tektites *Tektites found in the Ivory Coast region of Africa. They seem to date from the Lower Pleistocene Period.

J

Jansky, Karl Guthe (1905-1949) American engineer of Czech descent. In 1931, while investigating 'static' by using an improvised aerial, he detected radio emissions from the Milky Way. This was the start of *radio astronomy, but Jansky never really followed it up, and after 1937 he paid little further attention to it.

Janssen, Pierre Jules César (1824-1907) French astronomer, who became Director of the *Meudon Observatory in 1876. His main research was in connection with the Sun, and in 1868, independently of *Lockyer, he discovered the method of observing *prominences without waiting for an eclipse.

Janus A small satellite of Saturn; for data, see *Satellites. It is co-orbital with *Epimetheus. In 1966 A. Dollfus reported the discovery of an inner satellite, and it was named Janus. (I also saw it, using a 10-in refractor, but I did not recognize it as being new, and so can claim absolutely no credit!) Dollfus' 'Janus' was not confirmed until 1978, when it and Epimetheus were discovered; no doubt Dollfus had seen them both.

Jeans, Sir James Hopwood (1877-1946) British astronomer, noted for his work in astrophysics and cosmology as well as for his popular books and broadcasts.

Jet Propulsion Laboratory The centre for studies and tracking of planetary spacecraft, at Pasadena in California.

Jewel Box The nickname for the lovely star-cluster round *Kappa Crucis, in the *Southern Cross.

Jodrell Bank The site of the great radio astronomy observatory at Lower Withington, near Macclesfield in Cheshire. The main instruments are the 250-ft and 210-ft 'dishes'. The founder and the first director of the observatory was Sir Bernard *Lovell. The 250-ft 'dish' was the first large instrument of its type,

and was responsible for major advances. It was also used to track the first *artificial satellite, Sputnik 1.

Jones, Sir Harold Spencer (1890-1960) Astronomer Royal between 1933 and 1955. He was a Cambridge graduate, and had previously been HM Astronomer at the Cape. He was concerned largely with star catalogues and stellar radial velocities, but he also made a redetermination of the length of the *astronomical unit by observations of *Eros—though the value which he gave has proved to be slightly too great. He was a great administrator, who supervised the move of the Royal Observatory from Greenwich Park to Herstmonceux.

Jodrell Bank radio telescope.

Comparative sizes of Jupiter and the Earth.

Jovian Planets A common name for the giant planets Jupiter, Saturn, Uranus and Neptune.

Joy, James Harrison (1882-1973) American astronomer, noted mainly for his outstanding work in connection with stellar distances, variable stars, and stellar radial motions. He worked at the *Yerkes Observatory and then at *Mount Wilson, retiring in 1952.

Julian day A count of the days, reckoning from 12 noon on 1 January BC 4713—a starting-point chosen quite arbitrarily by the mathematician Scaliger, who introduced the system in 1582. The 'Julian' is in honour of Scaliger's father, Julius, and has nothing to do with Julius Cæsar.

Belts and zones of Jupiter.

Julian dates are widely used by variable star observers.

Juno Asteroid No 3, discovered by K. Harding (one of the so-called *Celestial Police, and assistant to *Schröter) in 1804. It is somewhat irregular in shape, with a diameter of 143 miles by 106 miles, and is thus much the smallest of the four original asteroids. For data see *Minor Planets. The mean opposition magnitude is 8.7, so that Juno is always well below naked-eye visibility.

Jupiter The largest planet in the Solar System. For data, see *Planets.

Jupiter is more massive than all the other planets combined (it has only 1/1047 times the mass of the Sun). Its equatorial diameter is over 88,000 miles, but its polar diameter is less than 84,000 miles, because the globe is obviously flattened. This is because of Jupiter's quick rotation, which makes the equator bulge out. The flattening is much greater than in the case of the Earth, which amounts to only 26 miles. This is partly because Jupiter is spinning more quickly, and partly because it is not a solid, rocky body. According to modern theory there is a rocky core made up of iron and silicates, at a temperature of about 30,000°C; around this is a thick shell of liquid metallic hydrogen, which is overlaid by another thick shell, this time of liquid molecular hydrogen. Above this comes

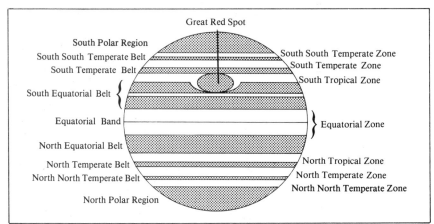

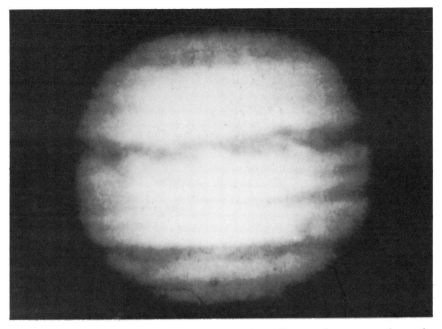

Jupiter.

the 'atmosphere', about 600 miles deep, which is made up of 81 per cent hydrogen, 17 per cent helium and 1 per cent of other elements. The atmosphere contains water droplets, ice crystals, ammonia crystals and ammonium hydrosulphite crystals.

Telescopically Jupiter shows a yellowish disk, crossed by the famous cloud belts. Gases warmed by the internal heat of the planet rise into the upper atmosphere and cool, forming clouds of ammonia crystals floating in the gaseous hydrogen. These clouds make up the bright zones, which are colder and higher than the dark belts. There are varied colours, due to the characteristic Jovian chemistry.

Jupiter does not rotate as a solid body would do; the rotation is *differential. The equatorial region (known as System I) has a period of 9 hours 50 minutes 30 seconds; it is bounded by the north edge of the south equatorial belt and the south edge of the north equatorial belt. The rest of the planet (System II) has a period of 9 hours 55 minutes 41 seconds, though various discrete features, such as the spots, have periods of their own, and move around in longitude.

Generally there are several belts visible with a small telescope. The north equatorial belt is always present, and is usually (though not always) the darkest and broadest of the belts; the south equatorial belt is more variable, and there have been rare occasions when the equatorial belts have merged. The south temperate belt is also prominent at times, and never disappears. The nomenclature of the various belts and zones is shown in the diagram.

Spots are common, though often short-lived. The most famous feature is the Great Red Spot, which has been under observation since the 17th century, though it sometimes disappears temporarily; it encroaches into the southern part of the south equatorial belt, forming what is called the Red Spot Hollow. It was once believed to be a semi-solid body floating in the outer gas, or else the top of a stagnant gas-column, but it is now known to be a whirling storm—a phenomenon of Jovian meteorology. At its largest it has a length of 28,000 miles and a maximum breadth of 8,000 miles, so that its surface area is greater than that of the Earth. It became very prominent in 1878, and was then of a brick-red colour. There is some evidence that it is decreasing in size, though it will certainly persist for a long time yet. The colour, still very pronounced on most occasions, may be due to phosphorus.

Four space-craft have now by-passed Jupiter: *Pioneer 10 (1973), *Pioneer 11 (1974) and *Voyagers 1 and 2 (1979). These probes have revolutionized all our knowledge of the planet, and have sent back detailed pictures as well as a wealth of information of all kinds. Jupiter was already known to be a source of radio waves; the probes confirmed that there is an immensely strong magnetic field, and that there are powerful radiation belts which would be instantly lethal to any astronaut unwise enough to enter them. Jupiter's magnetic field is of reverse polarity to ours (a compass needle there would point south), and the 'magnetotail' is so long that it can extend as far as the orbit of Saturn.

Jupiter has a dark ring, discovered by the space-craft but not observable from Earth; it may be less than half a mile thick, and is quite unlike the bright ring-system of Saturn. Auroræ on the night side of the planet were also recorded by the Voyagers.

Jupiter sends out more energy than it would do if it depended entirely upon what it receives from the Sun. It was once believed that this excess energy was due to the slow shrinkage of the planet, but it now seems more likely that the excess is simply the heat 'left over', so to speak, from the time when Jupiter was formed (Jupiter is now believed to be mainly liquid, and liquids are incompressible.) Below the outer clouds there must be a region where the temperature is comparable with that of the Earth, but if life exists there, as has been suggested, it must be very different from any form we know, and the idea seems rather far-fetched. We should learn more from the *Galileo satellite, due for launching in the near future. There are sixteen known satellites, of which the most important are the four *Galileans.

K

Kaiser Sea An obsolete name for the *Syrtis Major, the most prominent dark marking on Mars. The name honoured the Dutch astronomer F. Kaiser (1808-1872), a famous planetary observer.

Kant, Immanuel (1724-1804) German philosopher, remembered mainly for proposing a theory of the origin of the *Solar System from a rotating gas-cloud.

Kappa Crucis The Jewel Box cluster in the *Southern Cross—one of the most beautiful in the sky. There are many hot, bluish stars, with one prominent red supergiant about 80,000 times as luminous as the Sun. The distance is about 7,700 light-years. The cluster, known officially as NGC 4755, is relatively young. Unfortunately it is too far south to be seen from Britain or most of the United States.

Kappa Cygnids A minor *meteor shower. The maximum falls on 20 August, but the usual *ZHR is only about 4.

Kappa Pavonis A Type II Cepheid or *W Virginis variable, with a period of 9.1 days and a magnitude range of from 3.9 to 4.8. It lies in the far south of the sky, at a *declination of $-71\frac{1}{2}°$.

Kapteyn, Cornelius (1851-1922) Dutch astronomer and cosmologist who drew attention to the phenomenon of *star-streaming and was concerned largely with stellar proper motions. The 40-in reflector at *La Palma is now known as the Kapteyn Telescope.

Kellner eyepiece A type of *eyepiece using one convex lens together with a plano-convex lens.

Kepler, Johannes (1571-1630) Great German mathematician, who used the observations made by *Tycho Brahe to show that the planets move round the Sun not in circles, but in elliptical orbits. He also improved telescope designs, and

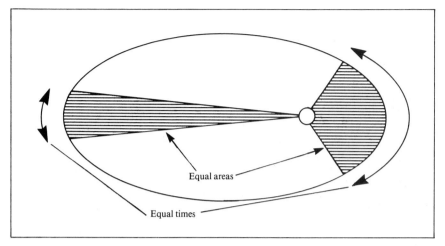

Kepler's Second Law.

made many other contributions to mathematical astronomy.

Kepler's Laws Three important laws of planetary motion, announced by *Kepler between 1609 and 1618. They are as follows:

1. The planets move in elliptical orbits, the Sun being situated at one focus of the ellipse while the other focus is empty.

2. The radius vector, or imaginary line joining the centre of the planet to the centre of the Sun, sweeps out equal areas in equal times. (In other words, a planet moves at its fastest when it is at its closest to the Sun.)

3. The squares of the *sidereal periods of the planets are proportional to the cubes of their mean distance from the Sun.

It was the announcement of these laws, following years of careful work by Kepler, which really disproved the old *geocentric or Ptolemaic theory of the universe. Kepler's Laws also apply, of course, to bodies such as comets, and to satellites moving round their primary planets.

Kepler's Star The usual name for the *supernova of 1604, in Ophiuchus.

Keyhole Nebula The dark nebulosity associated with *Eta Carinæ. Its nick-

name comes from its rather obvious shape.

Kiloparsec One thousand *parsecs (3,260 *light-years).

Kirchhoff's Laws of Spectral Analysis These were laid down by G. Kirchhoff in the early 1860s. They are as follows:

1. An incandescent solid, liquid or high-pressure gas emits a *continuous* spectrum—a coloured rainbow from red at the long-wave end through to violet at the short-wave end.

2. An incandescent gas at lower pressure emits an *emission* spectrum, made up of isolated bright lines each of which is characteristic of a particular element or group of elements.

3. When a continuous spectrum is viewed through lower-pressure gases, the bright lines are *reversed*, and appear dark (*absorption lines). The absorption lines in the spectrum of the Sun are often known as *Fraunhofer lines.

Kitt Peak Observatory America's national observatory, in Arizona. The main instruments are the 158 in *Mayall reflector and the *McMath solar telescope, but there are many other telescopes and instruments of all kinds.

Kleinemann-Low Object (KL) Infra-red

The McMath solar telescope at Kitt Peak.

source deep in the *Orion nebula. Its exact nature is still uncertain.

Kocab The star Beta Ursæ Minoris. It is of magnitude 2.1, and has a K-type spectrum, so that it is obviously orange. Its distance is 94 light-years, and it is about 100 times as luminous as the Sun. Apart from *Polaris it is the only bright star in Ursa Minor.

Kohoutek's Comet The comet of 1973. It did not become so brilliant as had been expected, but it was scientifically very useful, as it was studied by the astronauts aboard the orbiting space-station *Skylab, and was found to be surrounded by a huge cloud of rarefied hydrogen.

Konkoly Observatory Leading Hungarian observatory, and a centre for variable star research. It was founded by N. von Konkoly (1842-1916) who in 1898 presented it to the Hungarian Government.

Kopff's Comet A periodical comet, discovered in 1906; it has been seen at every return since 1919. The period is 6.4 years. There are plans to send a space-probe to it in the early 1990s.

Kordylewski clouds See *Lagrangian points.

Kreep A lunar basalt, characterized by its high content of potassium (K), rare earth elements (REE) and phosphorus (P).

Krüger 60 A binary star, 12.8 light-years away, near Delta Cephei. Each component is an M-type red dwarf; the fainter component, Krüger 60B (otherwise known as DO Cephei) is a *flare star, and is of exceptionally low mass, with a luminosity only 0.004 that of the Sun. The real separation between the components is of the order of 850 million miles, less than that between the Sun and Saturn. Relative motion can be detected over a period of only a few years, since the revolution period is only 44½ years.

Kuiper, Gerard Peter (1905-1973) Dutch astronomer, who spent most of his career in America and was Director of the *Jet Propulsion Laboratory from 1960 until his death. He founded the Lunar and Planetary Laboratory at Tucson, Arizona. He made major contributions to lunar and planetary research, particularly in connection with space-craft, and the first crater to be identified on Mercury, from *Mariner 10, was named in his honour. It was Kuiper who recommended the summit of Mauna Kea, in Hawaii, as a site for a major observatory.

Kuiper Airborne Observatory The world's first aerial observatory—a converted aircraft carrying a powerful telescope. The aircraft is a Lockheed C-141, and the 36-in reflector is installed in an open cavity recessed into the port side, immediately ahead of the wing. At peak altitude it is above 84 per cent of the Earth's atmosphere.

L

La Cilla Observatory One of the major observatories in Chile. The largest telescope is the 150-in reflector, completed in 1975.

Lacaille, Nicolaus Louis de (1713-1762) French astronomer, remembered chiefly for surveying 10,000 southern stars during his visit to the Cape from 1750 to 1754; he compiled a catalogue of 2,000 of these stars, and discovered many clusters and nebulæ.

Lagoon Nebula Messier 8 (NGC 6523). Emission nebula and galactic cluster in Sagittarius. It is just visible with the naked eye, and when photographed with large telescopes it is a glorious sight, with prominent dark 'lanes'.

Lagrange, Joseph Louis de (1736-1813) Great French mathematician (born in Italy), who was one of the pioneers of modern dynamical astronomy.

Lagrangian points Positions in space where a body of low mass can maintain a stable orbit in spite of the gravitational effects of more massive bodies. The two groups of *Trojan asteroids move at Lagrangian points in Jupiter's orbit, one group 60° ahead of Jupiter and the other 60° behind. Dim 'clouds' have been suspected in the Lagrangian points 60° ahead and 60° behind the Moon in its orbit, by the Polish astronomer K. Kordylewski, but have never been confirmed. Small satellites move in the Lagrangian points of several planetary satellites, notably *Tethys and *Dione in Saturn's system.

Lalande, Joseph Jérôme (1732-1807) French astronomer and mathematician, who became Director of the Paris Observatory. In addition to his scientific work he was also a well-known popularizer of astronomy.

The Lagoon Nebula (Steward Observatory, University of Arizona).

Laplace, Pierre Simon (1749-1827) French astronomer and mathematician. In 1796 he published his *Système du Monde*, containing his *Nebular Hypothesis of the origin of the Solar System—which was accepted for many years.

Las Campanas Observatory A major observatory near La Serena in Chile, operated from the United States. The chief instrument is the 100-in Irénée du Pont reflector.

Woodcut depicting the Leonid Meteor Storm in 1833.

Lassell, William (1799-1880) English amateur astronomer. By profession he was a brewer, but devoted all his leisure time to astronomy. He made his own telescopes, and with his 24-in reflector he discovered *Triton, the main satellite of Neptune, in 1846; he was also an independent discoverer of Saturn's satellite *Hyperion, and in 1851 he detected two of Uranus' satellites, *Ariel and *Umbriel. For some years he observed from the clearer skies of Malta.

Late-type stars Conventionally, stars of spectral types M, R, N and S.

Latitude, Celestial The angular distance of a celestial body from the nearest point on the *ecliptic.

Le Monnier, Pierre Charles (1715-1799) French astronomer, who observed *Uranus several times without recognizing it as a planet. He carried out much useful work, but it is also said that he never failed to quarrel with anyone whom he met!

Le Verrier, Urbain Jean Joseph (1811-1877) Brilliant French astronomer and mathematician, whose work led to the discovery of *Neptune in 1846. He also carried out outstanding work in connection with meteor streams. He became Director of the Paris Observatory, but his unpopularity forced his resignation in 1870, though he was subsequently reinstated.

Leavitt, Henrietta Swan (1888-1921) American woman astronomer. She discovered 2,400 variable stars, four novæ and several minor planets, but is remembered for her discovery, in 1912, of the *Period-Luminosity Law of *Cepheids.

Leda The 13th satellite of Jupiter. For data, see *Satellites.

Leiden Observatory The leading observatory of Holland; a centre for both optical and radio studies.

Lemaître, Georges (1894-1966) Belgian

priest, who was also a leading mathematician. In 1927 he published an important paper dealing with the *Big Bang theory of the universe.

Leonids A meteor shower, reaching its maximum on 17 November. In most years the Leonids are sparse, but occasionally they can give brilliant displays, as in 1799, 1833, 1866 and 1966. They are associated with the periodical comet *Tempel-Tuttle.

Libration Though the Moon always keeps the same face turned toward the Earth, because its rotation is *captured, we can examine a total of 59 per cent of the surface at various times; only the remaining 41 per cent is permanently averted. The fact that from Earth we can see more than half the surface, though of course never more than 50 per cent at any one moment, is due to effects known as librations.

Though the Moon rotates on its axis at a constant rate, it does not move in its orbit at constant velocity, because its path is not circular—and following *Kepler's Laws, it moves quickest when closest to us. This means that the amount of rotation and the position in orbit become periodically 'out of step', so to speak, and the Moon seems to oscillate slightly; first a little of the eastern limb is exposed and then, a few days later, a little of the western. This is *libration in longitude*. There is also a *libration in latitude*, due to the fact that the Moon's equator is tilted to the plane of its orbit by over 6°; and thirdly there is a *diurnal libration*, because the Earth itself is spinning, taking the observer with it. When the Moon is on the horizon, the observer is 'elevated' above the centre of the Earth by about 4,000 miles (the Earth's radius), so that he can see for an extra one degree round the mean limb of the Moon.

Lick Observatory Major American observatory, on Mount Hamilton in California. Its main instrument is a 120-in reflector; there is also one of the world's largest refractors, the 36-in, completed in 1888.

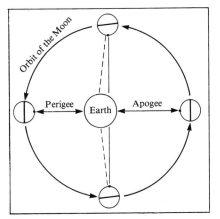

Libration. The Moon moves quickest at perigee, slowest at apogee, but the amount of axial rotation is constant so that the Moon appears to tilt very slowly. This is libration in longitude.

Life on other worlds Nobody is sure how life on Earth began, but at any rate we can lay down some conditions which are necessary for life as we know it. There must be an atmosphere, containing sufficient oxygen; there must be water, and there must be a suitable temperature-range. In the Solar System, only the Earth seems to fulfil all these requirements.

Most of the planets and satellites can be ruled out at once. There is no trace of life, either past or present, on the *Moon; the rocks which have been brought back are completely sterile. *Mercury has virtually no atmosphere, and *Venus is as hostile as it could possibly be. The giant planets have gaseous surfaces, and *Pluto is hopelessly cold. Among the satellites, only *Titan has an appreciable atmosphere so far detected, and Titan is unsuited in other ways. *Mars is the least hostile of all the planets, but the carbon-dioxide atmosphere is very thin, and the *Viking probes have shown no signs of living matter there.

Yet it must remembered that the Sun is an ordinary star, and few modern astronomers doubt that other stars have planet-systems of their own. There must surely be many millions of 'other Earths' capable of supporting life, and it is likely that intelligence is spread widely through the universe, though we cannot tell what form it may take. Direct contact will be very difficult indeed, and at present the only hope of communication seems to be by radio, though all experiments so far—such as *Ozma—have given negative results.

Of course, it is always possible that there are alien life-forms which can exist under conditions totally unsuited to Earth-type beings, but all we know about science argues against anything of the sort.

Light, Velocity of Light moves at a rate of 186,000 miles or 300,000 kilometres per second. It takes light only 8.6 minutes to reach us from the Sun, but over four years from the nearest star beyond the Solar System, *Proxima Centauri.

Light-curve A graph showing the changing brightness of a variable star. The magnitude of the star is plotted against the period, as shown in the diagram. In some cases the light-curve is regular, as with the *Cepheids, while for other types of stars the curves may be quite irregular.

Light-year The distance travelled by light in one year. It is equal to about 5,880,000 million miles or 9,460,000 million km.

Limb darkening The decrease in the intensity of light coming from the limb of a body such as the Sun—because light coming from the centre of the disk has to travel through a shallower layer of the solar atmosphere, and is less absorbed.

Lindblad, Bertil (1895-1965) Swedish astronomer, who became Director of the Stockholm Observatory in 1927. He was an expert in celestial dynamics, and made many important contributions, notably with regard to the rotation of the *Galaxy.

Linné Small feature on the lunar Mare *Serenitatis. In 1866 the German astronomer J. *Schmidt announced that Linné had changed from a crater into a white patch, but few astronomers now believe that any real change occurred there. The modern form of Linné, as photographed from space-probes, is that of a small, well-marked crater.

Lithosphere The solid upper crust of the Earth.

Lobate Scarps Cliffs seen on *Mercury, sometimes well over a mile high and 300 miles long. Nothing quite like them is found on the Moon or Mars.

Local Group A group of *galaxies of which our own *Galaxy is a member. The group contains three spirals (our Galaxy, M.31 in Andromeda and M.33 in Triangulum), the two *Magellanic Clouds, and over two dozen dwarf systems, as well as—probably—the massive elliptical system *Maffei 1, which is hard to observe because it is so heavily obscured by matter in the plane of the Milky Way.

Lockyer, Sir Norman (1823-1920) Pioneer English astrophysicist and spectroscopist, who, independently of *Janssen, discovered the method of observing solar *prominences without waiting for an eclipse.

Lohrmann, Wilhelm Gotthelf (1789-1840) German land surveyor and amateur astronomer. He began a large-scale lunar map of high quality, but his health failed before he had been able to complete it. The later map by J. *Schmidt was based upon it.

Loki A violently active sulphur volcano on *Io.

Light curve of δ Cephei.

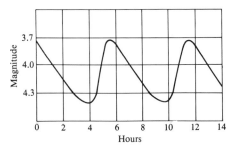

Above *Viking Lander 1 taking soil samples on Mars.*

Below *The north polar region of Mars.*

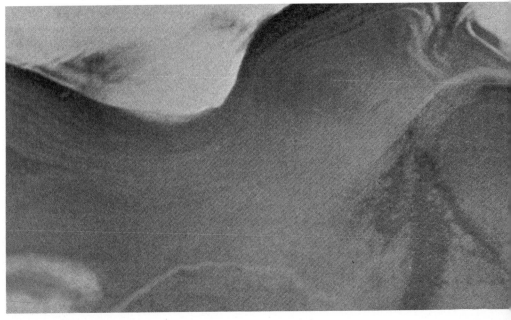

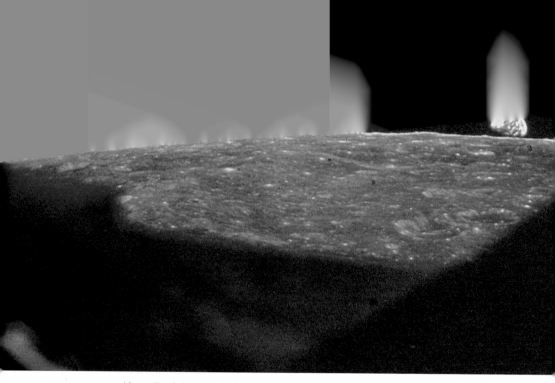

Above *Earthrise over the Moon, seen from an Apollo capsule.*

Below *A similar view photographed from Apollo 17.*

Above *The beautiful planetary nebula in Aquila, NGC 6781.*

Below *The Veil Nebula in Cygnus, remnant of a supernova over 50,000 years ago.*

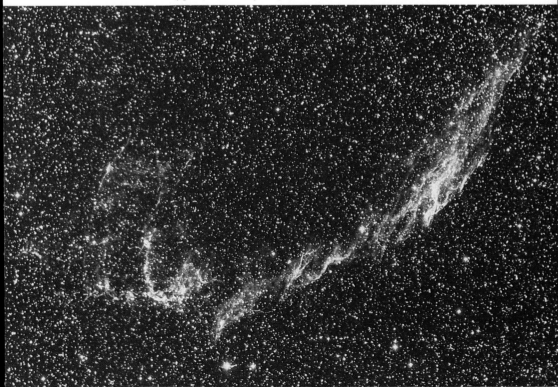

Orange soil on the Moon, photographed by the crew of Apollo 17.

Launch of the Giotto probe to investigate Halley's Comet.

Lomonosov, Mikhail (1711-1765) The first great Russian astronomer—also a chemist, poet and grammarian. In 1761 he rightly concluded that *Venus has a dense atmosphere; but his most important contribution was his championship of the *Copernican and Newtonian theories, which at that time were not widely accepted in Russia.

Longitude, Celestial The angular distance of a body from the *vernal equinox or First Point of Aries, measured eastward along the *ecliptic from 0° to 360°.

Lovell, Sir (Alfred Charles) Bernard (1913-) Great British radio astronomer, who was responsible for the setting-up of the 250-ft 'dish' at *Jodrell Bank and was the first Director of the Jodrell Bank Observatory.

Lowell, Percival (1855-1916) American astronomer, who founded the Lowell Observatory at Flagstaff, Arizona, and whose calculations led to the later discovery of *Pluto (though whether or not this was fortuitous is a matter for debate). Lowell was an expert mathematician, a benefactor of astronomy, and a brilliant writer and speaker, though, rather unfairly, he is best remembered today for his admittedly wild theories about artificial canals on Mars.

Luna probes The first Russian space-craft to the Moon, beginning in 1959 with Luna 1 (which by-passed the Moon), Luna 2 (which crash-landed there) and Luna 3 (which, in October, went round the Moon and sent back the first pictures of the far side, which is always turned away from the Earth). Altogether 24 Luna probes have been sent up; Lunas 16 and 20 brought back samples from the Moon, while Lunas 17 and 21 carried the *Lunokhods.

Lunation The interval between one new moon and the next. It is equal to 29 days 12 hours 44 minutes. An alternative name for it is the *synodical month*.

Lunokhods 'Crawlers' taken to the Moon, which operated for months,

Dome of the 24-in refractor at the Lowell Observatory.

sending back data as well as panoramic pictures. There were two: Lunokhod 1 (1970, carried by Luna 17) and Lunokhod 2 (1973, carried by Luna 21). Both were extremely successful.

Lyot, Bernard (1897-1952) French astronomer. He invented the *coronagraph and the *Lyot filter. All his career was spent at the *Meudon Observatory, Paris. He died suddenly while returning from an eclipse expedition.

Lyot filter A device used for observing solar *prominences at times of non-eclipse, together with other features of the solar atmosphere. It may also be called a *monochromatic filter*.

Lyrid Meteors An annual meteor shower,

105

reaching maximum on 22 April. The usual *ZHR is about 15. The Lyrids are connected with Comet Thatcher, last seen in 1861.

Lysithea The tenth satellite of Jupiter. For data, see *Satellites.

M

Mädler, J. H. von (1794-1874) German astronomer, who, with *W. Beer, drew up the best lunar map of its time; it was published in 1837-8, together with a complete description of the Moon's surface. Mädler subsequently left Berlin to become Director of the Dorpat Observatory in Estonia.

Maffei Galaxies Two galaxies discovered by Paolo Maffei in 1968; they lie in

The Small Cloud of Magellan.

Cassiopeia, and are so heavily obscured by material in the plane of our Galaxy that we cannot obtain a proper view of them. Maffei 1 is an elliptical system, probably a member of the *Local Group, but Maffei 2, a spiral, is probably too far away to belong to the Local Group.

Magellanic Clouds (Otherwise known as the Nubeculæ, or Clouds of Magellan.) Two important companions of our *Galaxy. Both are easy naked-eye objects, though too far south to be seen from Europe; the Large Cloud lies mainly in the constellation of Dorado, and the Small Cloud in Tucana. Superficially they look like detached portions of the Milky Way, but in fact the Large Cloud is 160,000 light-years away and the Small Cloud 190,000 light-years.

The centres of the Clouds are only 75,000 light-years apart, and the systems are enclosed in a common envelope of very rarefied hydrogen; there is also an 'arm' linking in this with our Galaxy, and there may even be an arm projecting on the far side of the Galaxy. It has been suggested that the Clouds are genuine satellites of our Galaxy, though on this point opinions differ. They are more or less irregular in form, though traces of spirality have been claimed for the Large Cloud, and there have even been vague suggestions that the Small Cloud may be a double system, with one part of it lying behind the other. The Large Cloud is about one-quarter the size of our Galaxy, and the Small Cloud one-sixth the size of our Galaxy.

The Clouds contain objects of all kinds; open and galactic clusters, gaseous nebulæ, stars of every type, radio sources, and even pulsars. They therefore present an excellent sample of the objects also found in our Galaxy, with the advantage that all these objects are at approximately the same distance from us. For example, it was by studying the *Cepheid variables in the Small Cloud that Miss *Leavitt, in 1913, made the discovery which led on to the Period-Luminosity Law. The Large Cloud contains the *Tarantula Nebula, the largest gaseous nebula known; if it were as near to us as the Orion Nebula, it would cast shadows. The exceptionally

The Large Cloud of Magellan (Photolabs Royal Observatory, Edinburgh, 1978).

luminous variable star *S Doradûs is also in the Large Cloud.

Since the Clouds lie at less than one-tenth the distance of the next nearest galaxies, the Andromeda and Triangulum spirals, they may be examined in considerable detail. Their importance to modern astronomers can hardly be over-estimated.

Maginus A large lunar crater; diameter 110 miles. It lies in the southern uplands, and becomes very obscure near full moon.

Magnetic storm A sudden disturbance of the Earth's magnetic field, shown by disturbances in radio communications as well as by fluctuations in the compass needle. It is due to charged particles sent out by the Sun, usually associated with solar *flares.

Magnetic variables Stars which have variable magnetic fields; the prototype is Alpha^2 Canum Venaticorum. The intensities of the spectral lines also vary.

Magnetosphere The area round a celestial body in which the magnetic field of that body is dominant. The Earth's magnetosphere includes the *Van Allen zones. The magnetosphere is shaped like a teardrop, with the tail pointing away from the Sun. On the sunward side of the Earth, the magnetosphere extends out to about 40,000 miles, but on the night side of the Earth is stretches for a much greater distance. As the *solar wind meets the magnetosphere it produces a shock-wave; inside the shock-wave there is a turbulent region, inside which is a definite boundary, the *magnetopause*. The magnetosphere proper lies on the Earthward side of the magnetopause. On the dark side of the Earth, the shock-wave merely weakens until it can no longer be detected.

The Moon, Venus and Mars have no detectable magnetic fields. That of Mer-

107

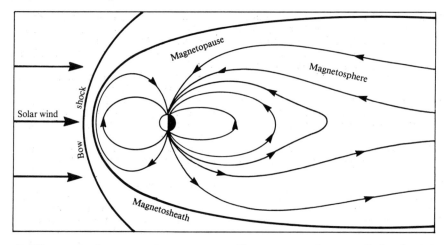

Earth's magnetosphere.

cury is appreciable, though much weaker than that of the Earth. Jupiter has an immense magnetosphere, and at times its 'magnetic tail' engulfs Saturn. Saturn's own field is strong, though by no means comparable with that of Jupiter.

Magnitude This is really a term for 'brightness', but astronomically there are several kinds of magnitudes.

Apparent or *visual magnitude* is the apparent brightness of a celestial body. The brighter the object, the lower the magnitude; thus Aldebaran in Taurus (magnitude 1) is brighter than Alphard in Hydra (magnitude 2). The faintest stars normally visible with the naked eye are of magnitude 6. The brightest stars have zero or, in a few cases, negative magnitudes; that of Sirius, the most brilliant star in the sky, is − 1.4. On the other end of the scale, the world's largest telescopes can now record objects down to about magnitude + 25. The magnitude scale is logarithmic; a star of magnitude 1 is a hundred times as bright as a star of magnitude 6. It is important to note that a star's apparent magnitude is no reliable key to its real luminosity; thus the Pole Star (magnitude 2) is more than three magnitudes fainter than Sirius, but it is much more remote, and is about 230 times as luminous.

On the stellar scale the brightest planet,

Venus, can reach a magnitude of − 4.4, while the full moon reaches − 12, and the Sun − 26.8.

Absolute magnitude is the apparent magnitude that a star would have if it were seen from a distance of 10 *parsecs, or 32.6 *light-years. At this distance, Sirius would have a magnitude of + 1.3, while that of Polaris would be − 4.6 — brighter than Venus appears to us. Absolute magnitude is, therefore, a measure of the star's real luminosity. The absolute magnitude of the Sun is + 4.8, so that from the standard distance it would be a dim naked-eye object.

Photographic magnitude has been described under the heading *Colour index.

Ordinary magnitudes are measured according to the amount of light received, but with a *bolometric magnitude* all the other wavelengths of the *electromagnetic spectrum are included as well, so that bolometric magnitude is a measure of the total amount of radiation being sent out by the star. The Sun's bolometric magnitude is + 4.6.

Maia One of the brighter stars of the *Pleiades cluster.

Main Sequence If the stars are plotted on a *Hertzsprung-Russell Diagram, according to luminosity and spectral type, it is found that most of them lie on a well-marked band from top left to bottom right, beginning with hot white or bluish stars (types O and B) and ending with feeble red stars (type M). This is the Main Sequence. The Sun, a yellow dwarf of type G, is a typical Main Sequence star. See also *Stars.

Maksutov telescope A special type of astronomical telescope, making use of both mirrors and lenses. It gives excellent results, and has a wide field of view. It was first described in detail in 1944 by the Russian astronomer after whom it is named.

Maraldi, Giacomo (1665-1729) Italian astronomer, nephew of G. D. *Cassini. He is remembered particularly for his observations of Mars.

Marduk A sulphur volcano on *Io. At the time of the Voyager 1 pass, the plume height was 75 miles.

Mare Latin for 'sea' (plural, maria).

Margaritifer Sinus (Gulf of Pearls) A prominent dark marking on Mars.

Marginis, Mare A small 'sea' close to the Moon's east limb.

Mariner probes American unmanned

Path of Mariner IV

spacecraft. Ten have been launched:
Mariner 1 (1962). Target: Venus. Complete failure at launch.
Mariner 2 (1962). Target: Venus. Successful.
Mariner 3 (1964). Target: Mars. Failed to achieve the right orbit.
Mariner 4 (1964). Target: Mars. Successful fly-by. Photographs taken.
Mariner 5 (1967). Target: Venus. Successful fly-by.
Mariners 6 and 7 (1969). Target: Mars. Successful photographic fly-bys.
Mariner 8 (1971). Target: Mars. Total failure at launch.
Mariner 9 (1971). Target: Mars. Entered Martian orbit, and sent back thousands of excellent photographs.
Mariner 10 (1973). Target: Mercury. The probe first by-passed Venus, and then made three successful photographic passes of Mercury.

Markab The star Alpha Pegasi, in the 'Square'. It is of magnitude 2.5; spectral type B9.

Markarian galaxies Unusual galaxies, to which attention was drawn by the Russian astronomer B. E. Markarian. They are very powerful in the ultra-violet region of the electromagnetic spectrum.

Mars The fourth planet in order of distance from the Sun. Data are given

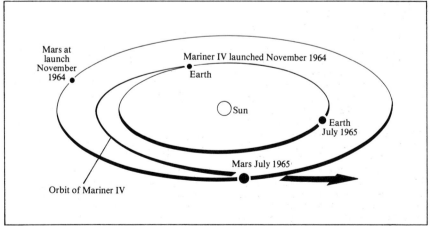

Mars at launch November 1964

Mariner IV launched November 1964

Earth

Sun

Earth July 1965

Mars July 1965

Orbit of Mariner IV

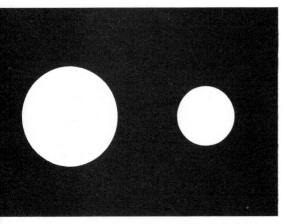

Comparative sizes of Mars and the Earth.

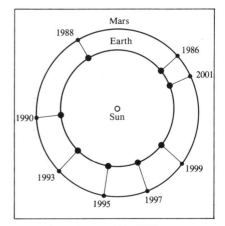

Oppositions of Mars, 1986-2001.

under the heading *Planets.

Mars was named after the mythological God of War, Mars (Ares) because of its strong red colour. When at its closest to us it may come within 35 million miles, and then outshines any other star or planet apart from Venus, but when at its faintest it sinks to the second magnitude, and is easily confused with a star. It shows per-

ceptible phases, first detected by *Galileo, and near *quadrature it appears strongly *gibbous. The *synodic period is 780 days, so that *oppositions do not occur every year; the next will be in July 1986, September 1988 and November 1990. Because the orbit of Mars is much more eccentric than ours, not all oppositions are equally favourable; that of 1988

Ice on the surface of Mars. It is extremely thin but may remain for approximately a hundred days. This photo was taken in the Utopia Planitia by Viking Lander 2 (JPL).

Part of the south polar cap region of Mars photographed by Mariner 7 (JPL).

Part of the Mare Sirenum on Mars photographed by Mariner 9.

will be particularly close, and Mars will attain an apparent diameter of almost 24 *seconds of arc.

Mars has always been regarded as the most Earthlike of the planets, despite its much smaller size; the escape velocity of 3.2 miles per second means than an atmosphere can be retained, though it has long been known to be much less dense than that of the Earth. Telescopically the planet shows a reddish disk, with white polar caps, and dark markings which are to all intents and purposes permanent; the most conspicuous of them, the *Syrtis Major, was seen by *Huygens as long ago as 1659. Originally it was thought that the dark areas were seas, but in the 19th century it became clear that this could not be so—the atmosphere is too thin—and

most astronomers came to the conclusion that the dark regions were old sea-beds filled with primitive 'vegetation'. The polar caps show seasonal changes; at maximum they are very striking, while in Martian summer they almost disappear. Since the tilt of Mars' axis is almost the same as ours, the seasons are of the same general type, though they are of course much longer.

Periodical dust-storms occur, sometimes veiling the entire surface. The famous *canals, drawn by *Lowell and others, were discredited well before the first space-probes by-passed the planet.

The *Mariner and *Viking probes have transformed all our ideas about Mars. The first successful space-craft, Mariner 4, flew past Mars in July 1965, and

111

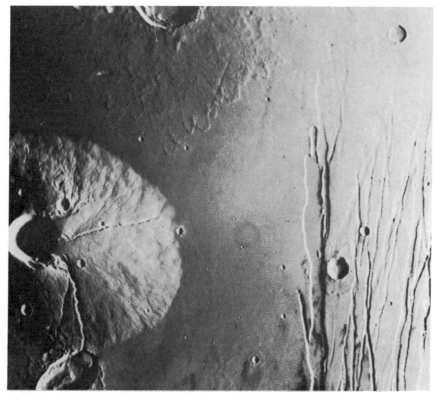

Above *Ceraunius Tholus, a Martian volcano, photographed by a Viking probe.*

Below *Drawing of Mars by Patrick Moore made on 5 December 1958 through a 6-in reflector.*

showed that the dark areas were not depressions; some, including the Syrtis Major, were elevated plateaux. Neither were there any signs of vegetation, and the atmosphere proved to be much thinner than had been expected; moreover it was made up mainly of carbon dioxide instead of the anticipated nitrogen. Mariners 6, 7, and 9 confirmed and amplified these findings. Then, in 1976, Vikings 1 and 2 made controlled landings, analyzing samples of Martian material and carrying out studies of the atmosphere. There was no trace of life in any form, either past or present, while the atmospheric density gave a pressure of below 10 millibars everywhere.

The orbiting sections of the probes have enabled us to draw up accurate maps of the whole surface. There are craters, ridges, mountains, valleys and lofty volcanoes, one of which (*Olympus Mons) rises to 15 miles above the outer surface level, and is topped by a huge, complicated caldera. Many of the features shown by the orbiters are almost certainly dry riverbeds, so that there must have been running water on Mars in the past.

The Russians have sent seven probes to Mars, but with scant success. There is no reason why a manned expedition should not be sent there in the foreseeable future, though when this will be attempted remains to be seen. It will certainly be preceded by a 'crawler' of the *Lunokhod type.

Mars has two small satellites, *Phobos and *Deimos, described separately.

Maskelyne, Nevil (1732-1811) The fifth Astronomer Royal. He was an able astronomer and administrator, but is best remembered because in 1767 he founded the *Nautical Almanac*.

Mass The quantity of matter that a body contains. It is not the same as weight; thus an astronaut on the Moon has only 1/6 his Earth weight, but his mass is unaltered.

Mass transfer With close *binary stars, it sometimes happens that one of the two components swells out, so that some of its mass can be transferred to the other

component. If the process continues for long enough, the originally less massive member of the pair may become the more massive of the two.

Mauna Kea Observatory The observatory on the summit of the 14,000 ft extinct volcano Mauna Kea, on Big Island of Hawaii. There are several large telescopes, and more are planned.

Maunder Minimum The period between 1645 and 1715, when there were very few sunspots. Attention was drawn to it by the British astronomer E. W. Maunder in the late 19th century.

Max Planck Institute for Radio Astronomy A major radio astronomy observatory in Effelsberg, near Bonn (West Germany), including the world's largest fully-steerable 'dish', 328 ft in diameter. It has been in operation since 1971.

Mayall telescope The 158-in reflector at the *Kitt Peak Observatory in Arizona.

Mayer, T. (1723-1762) German astronomer; a leading cartographer. Astronomically he is best remembered for his small but accurate map of the Moon.

McDonald Observatory American observatory on Mount Locke (7,000 ft high) in Texas. It has an 82-in reflector.

Map of Mars.

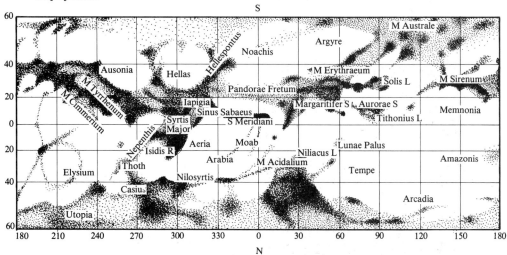

McMath-Hulbert Observatory Solar observatory in Michigan, USA.

McMath telescope The remarkable 'solar telescope' at *Kitt Peak. It is much the largest of its type in the world.

Mean sun An imaginary body travelling eastward along the celestial *equator, at a rate of motion equal to the average rate of the real Sun along the *ecliptic. Further details are given under the heading *Equation of Time.

Medii, Sinus (The Central Bay) A small bay near the apparent centre of the Moon's disk. It leads off the Mare *Nubium.

Megaparsec One million *parsecs.

Menzel, Donald Howard (1901-1976) American astronomer, who made many important advances in connection with solar research. He was also a noted author of popular books.

Merak The star Beta Ursæ Majoris; magnitude 2.37. It is the fainter 'Pointer' to the Pole Star (the other being *Dubhe). Merak is 30 times as luminous as the Sun, and is a white star of type A.

Mercury The first planet in order of distance from the Sun. For data, see *Planets.

Mercury is never prominent as seen with the naked eye, partly because it is small and partly because it stays in the same part of the sky as the Sun, so that it is never visible against a dark background.

Comparative sizes of Mercury and the Earth.

When at its best, it shows up as a bright starlike object in the west after sunset or in the east before sunrise. Rather surprisingly, the maximum magnitude is −1.9, brighter than any star—even Sirius.

Because Mercury is closer to the Sun then we are, it shows *phases similar to those of Venus, so that it may appear as a crescent, half or gibbous shape; when full it is on the far side of the Sun (at *superior conjunction) and is to all intents and purposes out of view. When it passes directly between the Sun and the Earth, it appears in *transit against the bright solar disk. The next transits are those of 13 November 1986, 6 November 1993 and 15 November 1999. (Transits can occur only in May or November.)

Very little can be seen on Mercury with Earth-based telescopes, and the best map of the pre-Space Age, compiled by E. M. *Antoniadi with the aid of the large 33-in refractor at *Meudon, has proved to be very inaccurate. Antoniadi, like *Schiaparelli before him, believed Mercury to have a *captured rotation (88 Earth-days), but it is now known that the real rotation period is 58.65 days, two-thirds of a Mercurian 'year'. To an observer on the surface of Mercury, the interval between one sunrise and the next would be 176 days or two Mercurian 'years'. Because Mercury's orbit is more eccentric than for any other planet (excluding Pluto) there are two 'hot poles', where the Sun is overhead when Mercury is at its nearest to the Sun.

Most of our knowledge of Mercury comes from one probe, *Mariner 10, which made three active passes of the planet: in March and September 1974 and March 1975. It was found that there is an excessively tenuous atmosphere and an appreciable magnetic field. The surface is very rough and crater-scarred; there are plains, valleys, ridges and scarps, together with what are termed 'intercrater plains' not like anything found on the Moon. Ray-craters also exist. The first of these to be identified, as Mariner 10 approached Mercury, was named Kuiper in honour of G. P. *Kuiper, who had played such a major rôle in planetary research. The most imposing formation is the Caloris

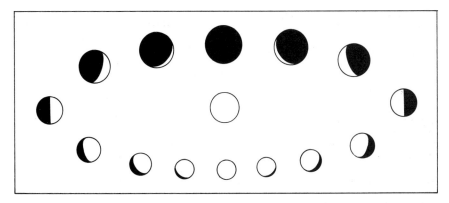

Above *Phases of Mercury and Venus.*   **Below** *Major features of Mercury.*

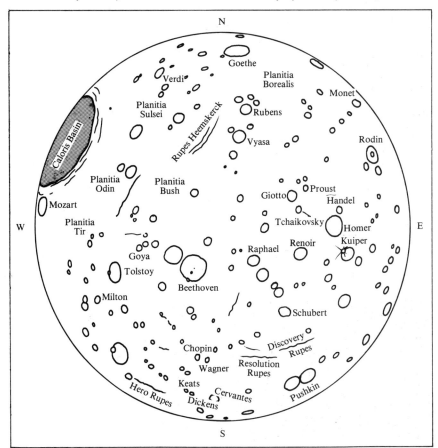

The Caloris Basin on Mercury photographed by Mariner 10.

Basin, over 800 miles in diameter—a vast ringed structure bounded by a ring of mountain blocks over a mile high in some cases. The Caloris Basin lies at one of Mercury's 'hot poles' (hence the name).

Unfortunately, at each pass of Mariner 10 the same part of Mercury was in sunlight, so that there is still a good part of the surface which is unmapped; we have charted only half of the Caloris Basin. But there is no reason to doubt that the unknown regions are essentially different from those surveyed by Mariner 10. The temperature range is very great, and any form of life there appears to be totally out of the question. Like Venus, Mercury has no satellite.

Mercury Programme The first US manned space-flights. There were seven successful missions, beginning with the sub-orbital 'hop' of Alan Shepard on 5 May 1961 and ending with the flight of Gordon Cooper in May 1963. In February 1962 John Glenn made the first American orbital flight, in Mercury 6 (better known as 'Friendship 7').

Meridian, Celestial The great circle on the *celestial sphere which passes through the *zenith and both celestial poles. It follows that the meridian cuts the observer's horizon at the exact north and south points. The *upper meridian* is the section from the south point up through the zenith to the north celestial pole. (This, of course, applies to the northern hemisphere of the Earth; from places south of the Earth's equator, the upper meridian will extend from the north point up through the zenith to the south celestial pole.)

Meridian circle A telescope which moves only in the north-south plane, and is used to make very accurate determinations of the altitudes—and, hence, the *declinations—of celestial bodies, mainly stars and planets.

Merope One of the brighter stars of the *Pleiades cluster.

Messier, Charles (1730-1817) French astronomer, interested mainly in comets. In his comet searches he was constantly misled by nebulæ and clusters, so that he decided to catalogue them as 'objects to avoid'. Ironically, it is for this catalogue that he is now best remembered. It was published in 1781.

Messier numbers The numbers allotted by *Messier to the clusters and nebulæ which he listed. The original catalogue contained 102 entries; five were added later. Messier's numbers are still widely used—thus the Andromeda Galaxy is M.91, the Orion Nebula is M.42, Præsepe is M.44, and so on. However, the M numbers have now been officially superseded by the *NGC numbers, given by *Dreyer in his catalogue published in 1886.

Messier's catalogue includes objects of all kinds; galaxies, gaseous nebulæ, and open and globular clusters; there is one supernova remnant, the *Crab Nebula (M.1). Obviously there are no M numbers for objects which are not visible from France.

Meteor A small particle, usually smaller than a grain of sand, moving freely around the Sun. When in space it cannot be seen, because it is too small, but when it dashes into the Earth's upper atmosphere, at a relative speed of anything up to 45 miles per second, it rubs against the air-particles and becomes heated by friction, destroying itself and producing the luminous effect known as a shooting-star. What we see, of course, is not the tiny particle itself, but the effects which it produces in the Earth's atmosphere as it burns away. Visual observations of meteors are still of value, though in modern times they are also tracked by *radar, since a radar pulse is reflected from a meteor trail.

Over 20 million meteors enter the atmosphere every day, but each is of very slight mass, and few penetrate to a height of below 40 miles above the ground without being destroyed. They are cometary débris, and therefore tend to travel in swarms. When the Earth passes through a swarm, the result is a shower of shooting-stars; the meteors of any particular shower will seem to come from one particular region of the sky, known as the *radiant, but this is merely an effect of perspective. A good way to explain it is to picture the scene from a bridge overlooking a motorway. The parallel traffic lanes seem to 'radiate' from a distant point near the observer's horizon. It is essentially the same with the meteors of a shower, which are travelling through space in parallel paths.

Some important annual showers are as follows: the Zenithal Hourly Rates or *ZHR values are not consistent, but act as a guide.

Shower	Date of maximum	ZHR	Parent comet
Quadrantids	3 January	110	
Lyrids	22 April	12	Thatcher (1861 I)
Eta Aquarids	4 May	20	Halley
Delta Aquarids	28 July		
Perseids	12 August	68	Swift-Tuttle
Draconids	10 October	variable	Giacobini-Zinner
Orionids	21 October	30	Halley
Taurids	4 November	12	Encke
Leonids	17 November	variable	Tempel-Tuttle
Andromedids	20 November	low	Biela
Geminids	14 December	58	
Ursids	22 December	12	Tuttle

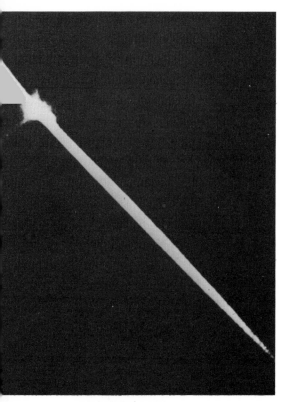

An exploding meteor.

The showers are named according to the constellation which contains the radiant (the Quadrantids come from the area of the now-rejected constellation of *Quadrans, near Boötes). The Perseids are the most consistent meteors, and always give a good display between about 28 July and 18 August; the Leonids are occasionally spectacular; the Andromedids, associated with the defunct *Biela's Comet, are now sparse. The Draconids can also be occasionally spectacular.

There are also *sporadic meteors*, not members of showers, which may appear from any direction at any moment. Really brilliant meteors are known as *fireballs. It is important to note that there is no association between meteors and *meteorites.

Meteor Crater An impact crater in Arizona, almost a mile in diameter, apparently formed about 22,000 years ago. It is the best example of an impact crater and is a well-known tourist attraction.

Meteorite A relatively large body which survives the complete drop to the ground without being destroyed, and may even produce a crater, such as that in Arizona. Meteorites are not connected with comets or with meteors; they are more nearly related to the asteroids or *minor planets—in fact there is probably no basic difference between a large meteorite and a small asteroid.

Meteorites are divided into several classes, according to their chemical composition. There are, for instance, *irons* or *siderites*; *stony-irons* or *siderolites*; and *stones* or *aerolites*. Among the aerolites, *chondrites* contain spherical particles called chondrules, which are mineral fragments with radiating structure; in *achondrites*, chondrules are absent. When etched with acid, irons will show the characteristic *Widmanstatten patterns, not found elsewhere. There is no record of any human death by a meteorite fall, though a few people have had narrow escapes.

The most massive known meteorite is still lying where it fell, in prehistoric times, at Hoba West, near Grootfontein in Southern Africa; it weighs over 60 tons. Other massive meteorites are the Ahnighito, from Greenland, now in the Hayden Planetarium, New York (30.4 tons), Bacuberito, Mexico (27 tons), Mbosi, Africa (26 tons) and Armanty, Outer Mongolia (about 20 tons). In 1947 there was a major fall in Siberia, in the Sikhote-Alin area, which produced over 100 small craters; the missile broke up during its descent. The 1908 missile, also in Siberia, was more probably part of a comet.

The most recent British falls have been those of 24 December 1965 (Barwell, Leicestershire) and 25 April 1969 (Bovedy, Northern Ireland). The *Barwell Meteorite was well observed during its descent, and many fragments were found. Most of the mass of the Bovedy Meteorite apparently fell in the sea.

Most museums have collections of meteorites, and new samples are recovered yearly; many, for example, have recently been found in Antarctica.

Metonic cycle A luni-solar cycle of 19 years, after which the phases of the Moon will recur on the same days of the year (allowing for the vagaries in our calendar). This is because 235 lunar months are almost equal to 19 solar years. It was discovered by the Greek astronomer Meton (*c* 432 BC) and provides a rough means of predicting eclipses; see also *Saros.

Meudon Observatory Major French observatory, near Paris. It contains the famous 33-in refractor, used by *Antoniadi for his planetary work.

Miaplacidus The star Beta Carinæ, too far south to be seen from Britain. For data, see *Stars.

Micrometeorites Extremely small particles, no more than 1/250 of an inch in diameter moving round the Sun. Their masses are too low for them to produce luminous effects when they enter the atmosphere, and they do not therefore produce shooting-star effects. Since 1957 they have been carefully studied by artificial satellites and space-probes.

Micrometer A measuring device, used together with a telescope to measure very small distances—such as the separations of the components of double stars. There are many forms, the most common being the *filar micrometer.

Micron A unit of length: 1/1000 of a millimetre (1/25,400 of an inch); there are 10,000 Ångströms to one micron. The usual symbol for a micron is the Greek letter mu (μ).

Microwaves Radiations intermediate between the infra-red and the radio part of the *electromagnetic spectrum: wavelengths from 1 millimetre to 1 metre.

*Debris at *Tunguska, Siberia, caused by the 1908 meteorite.*

Midas Asteroid No 1981; it is of the *Apollo type.

Midnight Sun The Sun seen above the horizon at midnight. This can occur during some part of the year anywhere within the Arctic or Antarctic Circle. At the North Pole, the Sun stays above the horizon all the time that it is north of the celestial equator, between about 21 March and 22 September; opposite conditions apply to the South Pole.

Mimas, one of Saturn's smaller satellites, photographed by Voyager 1 in 1980. The large Herschel crater is approximately 130 km in diameter (JPL).

Milky Way The luminous band stretching across the night sky. It has been known since the earliest times, and there are many legends about it, but not until the invention of the telescope was it found to be made up of stars. As explained under the heading *Galaxy, it is a line of sight effect; the stars in the Milky Way are not genuinely crowded together.

The term used to be applied to the Galaxy itself, but nowadays the name is restricted to the luminous band—though one still finds references to 'the Milky Way Galaxy'.

Milne, Edward Arthur (1896-1950) English astronomer, who made important contributions to astrophysics, and also developed a theory of 'kinematic relativity' which was for some time regarded as a serious rival to general Einsteinian *relativity.

Mimas The innermost satellite of Saturn known before modern times. For data, see *Satellites.

Mimas is only slightly denser than water, and has an icy surface; the globe is presumably made up of a mixture of rock and ice. The surface is dominated by a huge crater, *Herschel, which is one-third the diameter of Mimas itself, and has a massive central mountain; from wall to wall it measures 81 miles. There are many other craters, as well as ridges, hills and grooves. All our information about the surface features has been drawn from the results obtained from Voyagers 1 and 2. A very small satellite was found in 1982 moving in the same orbit as Mimas, but nothing definite is known about it.

Minkowski, Rudolf (1895-1976) German astronomer who spent most of his working life at Mount Wilson, in California. He concentrated upon supernovæ and studies of stellar evolution; he was also a pioneer of radio astronomy.

Minor planets Alternatively known as asteroids or planetoids. They are small worlds, most of which revolve round the Sun between the orbits of Mars and Jupiter. The largest of them, Ceres, has a diameter of 623 miles; of the rest, only

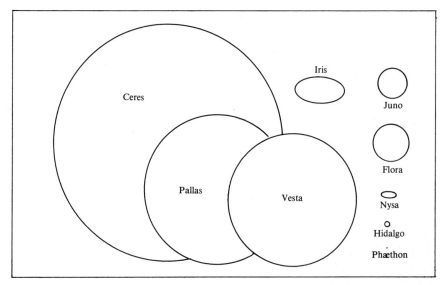

Comparative sizes of some asteroids.

Pallas, Vesta and Chiron exceed 300 miles, and only Vesta is ever visible with the naked eyes. Asteroids are of various types: carbonaceous, silicate, iron and so on. It was once believed that they were the débris of an old planet (or planets) which broke up, but it is now thought that they never combined into a larger body because of the powerful disruptive influence of Jupiter.

Asteroids are given permanent numbers when their orbits have been calculated with reasonable precision; the first four (discovered between 1801 and 1808) are 1 Ceres, 2 Pallas, 3 Juno and 4 Vesta. All the early names were mythological, but when these were exhausted other names were given—for instance 518 Halawe, after an Arabian sweet of which its discoverer (R. S. Dugan) was particularly fond! Since 1891 most discoveries have been made photographically.

Variations in magnitude allow the rotation periods of some asteroids to be found; they range from 2 hours 16 minutes for 1566 Icarus up to 1,500 hours for 280 Glauke. It is possible that some asteroids such as 532 Herculina are either very irregular in shape, or are double.

Data for the first ten known asteroids are as follows:

No	Name	Magnitude (mean opp)	Sidereal period in years	Mean distance from Sun in millions of miles	Diameter in miles
1	Ceres	7.4	4.60	257.0	623
2	Pallas	8.0	4.61	257.4	378
3	Juno	8.7	4.36	247.8	143
4	Vesta	6.5	3.63	219.3	334
5	Astraea	9.8	4.14	239.3	73
6	Hebe	8.3	3.78	225.2	121
7	Iris	7.8	3.68	221.5	130
8	Flora	8.7	3.27	204.4	94
9	Metis	9.1	3.69	221.7	94
10	Hygeia	10.2	5.59	292.6	280

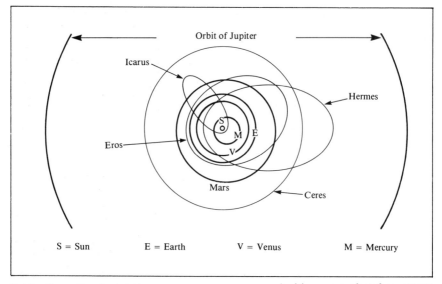

S = Sun   E = Earth   V = Venus   M = Mercury

Orbits of exceptional asteroids.

Occultations of stars by asteroids have given useful diameter measures. Not all asteroids are perfectly spherical; many of the smaller ones are not, and even Juno is decidedly irregular in shape.

Some asteroids depart from the main swarm. Such are the *Amor asteroids, whose orbits cross that of Mars; the *Apollo asteroids, which come within the Earth's orbit; the *Aten asteroids, whose orbits lie mainly within that of the Earth; and the *Trojans, which move in the same orbit as Jupiter. A few asteroids, such as 944 Hidalgo, have highly eccentric orbits. 2060 *Chiron, which moves mainly between the orbits of Saturn and Uranus, is of uncertain nature. Two known asteroids, 1566 Icarus and the recently-discovered *Phæthon approach the Sun within the orbit of Mercury.

Plans are now being made to by-pass asteroids with space-craft; the first target will be 29 Amphitrite, which will be surveyed by the *Galileo probe on its way to Jupiter.

The number of asteroids is certainly very great, and 40,000 is a conservative estimate. Of course, their low masses and escape velocities means that they cannot retain any trace of atmosphere.

Mintaka The star Delta Orionis, the northernmost of the three stars of Orion's Belt. It is of magnitude 2.2, but is very slightly variable; it is a *spectroscopic binary at least 20,000 times as luminous as the Sun. It is almost on the celestial equator; its declination is less than 18 minutes of arc south.

Minute of arc (symbol ′) One-sixtieth of a *degree of arc.

Mira Ceti The prototype long-period variable star. It was seen by D. Fabricius in 1596, though *Phocylides Holwarda in 1625, was the first to recognize it as a variable. Mira has a range of magnitude from 1.7 to 10, though on one occasion (in 1779) it is said to have attained 1.2; at other maxima it does not exceed magnitude 4. The mean period is 331 days, so that Mira is not a naked-eye object for more than a few weeks in every year. It has an M-type spectrum, and like all long-period variables of its type it is very red. The distance is 95 light-years according to a recent estimate. Mira has an easy telescopic companion of magnitude 10.

Miranda The innermost satellite of Uranus. For data, see *Satellites. It was surveyed from close range by the *Voyager 2 probe in January 1986.

Missing mass problem The movements of star-clusters, and of clusters of galaxies, seem to indicate that the systems contain much more mass than would be expected from the visible objects in them. The nature of this 'missing mass' is not yet understood.

Mizar The star Zeta Ursæ Majoris. It has a naked-eye companion, *Alcor. Mizar itself is a very easy telescopic double—the first to be recognized (by G. Riccioli, in 1651). Mizar A, the brighter component, was the first *spectroscopic binary to be recognized, by E. C. Pickering in 1889; Mizar B is also a spectroscopic binary; and to complete the picture Alcor too is a spectroscopic binary. The total luminosity of Mizar is about seventy times that of the Sun.

Molecule A stable association of atoms—or, broadly speaking, a group of atoms linked together. For example, a molecule of water is made up of two hydrogen atoms together with one oxygen atom (H_2O), while a molecule of carbon dioxide consists of one carbon atom together with two oxygen atoms (CO_2). Some molecules are highly complex. Many types of molecules have now been identified in space.

Molongo Radio Observatory An important Australian radio astronomy observatory at Hoskinstown in New South Wales.

Monochromatic filter A filter which allows the passage of light-waves only in a selected narrow region of the *electromagnetic spectrum. See *Lyot filter.

Month In everyday language, a month—more accurately, a *calendar month*—is either thirty or 31 days, apart from February, which has 28 days in ordinary years and 29 days in *leap-years. Astronomically, however, there are several kinds of 'months', all associated with the time taken for the Moon to complete one journey in its orbit.

The *anomalistic month* of 27.55 days is the time taken for the Moon to travel from one *perigee to the next.

The *sidereal month* (27.32 days) is the time taken for the Moon to go once round the Earth—or, more precisely, round the *barycentre—with respect to the stars.

The *synodical month* or *lunation* (29.53 days) is the time between successive new moons, or successive full moons.

The *nodical* or *Draconitic month* (27.21 days) is the time taken for the Moon to make successive passages through the same *node.

The *tropical month* (27.32 days) is the time taken for the Moon to return to the same *celestial longitude. It is only about 7 seconds shorter than the sidereal month.

Moon The Moon is generally called the Earth's satellite. For data, see *Satellites. However, the Moon has a mass of 1/81 that of the Earth, which makes it rather too considerable to be regarded as a mere satellite; it is probably better to class the Earth-Moon system as a double planet.

It used to be thought that the Moon broke away from the Earth in the remote past, perhaps leaving a depression now filled by the Pacific Ocean. This latter idea is completely out of court—the Pacific depression is insignificant compared with the size of the Moon—and in any case, almost all authorities agree that the Moon and the Earth have always been separate, though they are of the same age (about 4.7 thousand million years).

The Earth and Moon move round the *barycentre, or common centre of gravity of the Earth-Moon system, but since the

Comparative sizes of the Moon and the Earth.

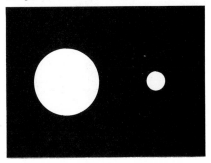

Moon

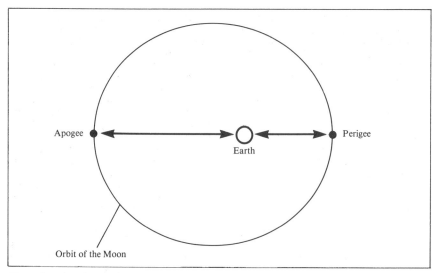

Above *Lunar perigee and apogee.*

Below *Lunar craters: Clasius to the upper right, Tycho lower right.*

barycentre lies inside the Earth's globe the plain statement that 'the Moon goes round the Earth' is good enough for most purposes (note, however, that the Moon's orbit is always concave to the Sun). Because it has no light of its own, the Moon shows phases from new to full. Its orbit is somewhat eccentric, ranging in distance from the Earth between 221,460 miles at *perigee out to 252,700 miles at *apogee. The apparent diameter varies between 29'22" and 33'31". The *albedo is low; on average only about 7 per cent. The surface gravity is only one-sixth that on Earth.

The Moon's surface contains broad, darkish plains or maria ('seas'), mountains, peaks, valleys, ridges, *rills, and vast numbers of walled formations known generally as craters. There have been endless discussions about the origin of the surface features. According to one theory, the craters are of internal origin ('volcanic', using the term in a very broad sense); alternatively, many astronomers believe that they have been formed by meteoritic impact. No doubt both processes have operated. Some of the craters are well over 150 miles across; others are tiny pits too small to be seen at all from Earth. The 'seas' never contained water, though at one time they were undoubtedly seas of lava.

Craters are named in honour of famous personalities—usually, though not always, astronomers. Some of the mountain ranges have 'terrestrial' names, such as the Alps and the Apennines; the main ranges form the borders of the regular maria.

The craters are not distributed at random; they form groups and chains—such as the great chain of formations near the apparent centre of the disk, of which

The Moon photographed through the 12-in refractor of Wilhelm-Foerster-Stenwarte in Berlin.

The crater Copernicus, one of the early lunar close-ups taken by Lunar Orbiter 2 in November 1966.

*Ptolemæus, over 90 miles in diameter, is the largest. Many craters have central peaks or mountain groups, but these peaks never attain the heights of the outer ramparts. When one crater interrupts another, it is almost always the larger formation which is broken, indicating greater age. There are many chains of small craters, and there are also domes— low swellings, often crowned by crater-lets.

Because the Moon has *captured rotation, it keeps the same face turned toward the Earth, though the effects known as *librations mean that altogether we can examine 59 per cent of the total surface at one time or another. The first pictures of the far side were obtained in 1959 by the Russian space-craft *Lunik 3. It was found that the far side is somewhat different from the near side; there are large light-floored craters or *thalassoids, but no major maria. Of special interest is the huge walled formation *Tsiolkovskii, with a dark floor and a central peak; it seems to be intermediate in type between a mare and a crater.

Man first reached the Moon in July 1969, when Neil *Armstrong stepped out from *Eagle*, the lunar module of Apollo 11, on to the barren rocks of the Mare *Tranquillitatis or Sea of Tranquillity. Five more successful expeditions followed, ending with Apollo 17 in December 1972. Altogether 2,196 samples were brought home, weighing in all 382 kilo-grammes. Recording stations or *Alseps were left behind, and were finally shut down on 30 September 1977.

The Apollo missions have improved our knowledge of the Moon beyond all recognition. There is an upper loose layer or *regolith, made up of fragmented material formed by the breaking-up of the underlying bedrock; its depth is variable from a few feet down to over 100 ft. The crust thickness is between 30 and 40 miles; then comes the mantle, which goes down to 600 miles or so; below this again comes the asthenosphere, a region of partial melting, and finally the core, with a diameter of something like 350 to 400 miles, which is fairly hot (perhaps as hot as 1,500° Centigrade). There is no detectable magnetic field, though various regions have localized magnetism.

All the rocks brought back by the Apollo astronauts (and the two Russian unmanned sample and return probes) are igneous; their ages range between 3.1 and 4.7 thousand million years. In the lavas, basalts are dominant; there are breccias in every part of the Moon so far studied. 'Moonquakes' have been recorded regularly, some shallow and some originating about midway between the Moon's crust and its centre. However, these tremors are not strong enough to threaten any future Lunar Base.

The Moon has been virtually inactive for at least a thousand million years, and probably longer. Minor traces of current

126

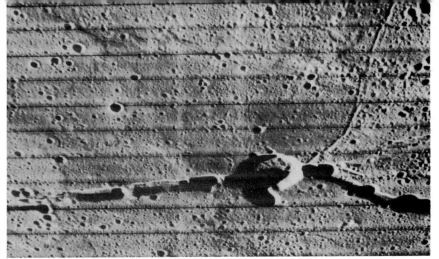

Above *View from Lunar Orbiter 3 of the crater Hyginus and two branches of the Hyginus Rill.*
Below *Main features of the near side of the Moon.*

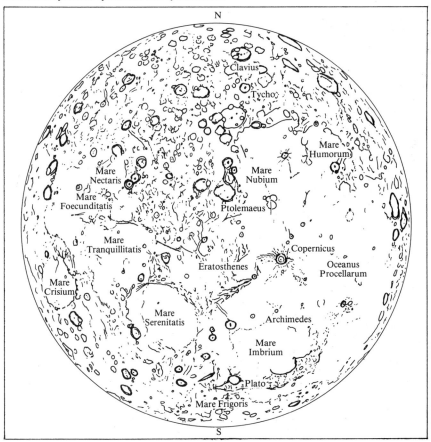

N

Clavius

Tycho

Mare
Humorum

Mare
Nectaris

Mare
Nubium

Mare
Foecunditatis

Ptolemaeus

Mare
Tranquillitatis

Copernicus

Eratosthenes

Oceanus
Procellarum

Mare
Crisium

Mare
Serenitatis

Archimedes

Mare
Imbrium

Plato

Mare Frigoris

S

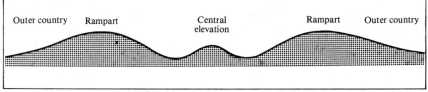

Cross-section through a typical lunar crater.

activity are known (see *Transient Lunar Phenomena), but these seem to be due to nothing more than mild gaseous escapes from beneath the regolith.

There is to all intents and purposes no lunar atmosphere, and the rocks show no signs of life either past or present. Yet despite its hostile nature, the Moon must now be regarded as accessible, and there is no reason why full-scale bases should not be set up there in the reasonably near future.

Moon Illusion When the Moon is low over the horizon it seems to look larger than when it is high up. This is pure illusion, since in reality the low Moon is no larger than the high Moon. *Ptolemy, about AD 150, wrote that an object seen across 'filled space' (such as the Moon near the horizon, seen across the Earth's surface) will give the impression of being more distant than an object seen across 'empty space' (such as the high Moon), so that it will seem to be larger. This may not be the full explanation, but in any case it is easy to prove, by a simple practical experiment, that the illusion *is* an illusion and nothing more.

Morning Star (Phosphorus) The old name for Venus as a morning object, in the east before dawn.

Mortis Lacus (The Lake of Death) It leads off the lunar Lacus *Somniorum, and includes an important system of *rills associated with the crater Bürg.

Mount Wilson Observatory American observatory in California, established in 1904 by G. E. *Hale. Of its main telescopes, the 60-in reflector and the 100-in *Hooker reflector are of special impor-

tance; until 1948 the Hooker telescope was the largest in the world. Unfortunately the growth of Los Angeles means that the conditions on Mount Wilson have deteriorated so badly that at the present moment it seems that the Observatory is likely to be phased out.

Moving clusters Groups of stars moving through space in the same direction at the same rate; thus five of the seven stars in the '*Plough' pattern of Ursa Major make up a moving cluster (the exceptions being *Alkaid and *Dubhe). If the velocities of the individual stars are known, and also the point in the sky to which they seem to be converging (because of perspective effects, since their paths are actually parallel), the distance of the group can be calculated. This has been done, for example, in the case of the *Hyades.

Moscoviense, Mare (The Moscow Sea) A small dark region on the Moon's far side. It was recorded by *Lunik 3, the first space-craft to obtain photographs of the averted regions.

Mount Stromlo Observatory Major Australian observatory, near Canberra. The largest telescope is a 74-in reflector.

Mu Cephei The *Garnet Star.

Müller, Johann See *Regiomontanus.

Multiple-Mirror Telescope (MMT) A reflecting telescope which uses several mirrors, working together. The first instrument of the kind was set up at Mount Hopkins, in Arizona, in 1978; it has six identical 72-in mirrors, and is equivalent to a single-mirror 176-in telescope. It has been very successful. A multiple-mirror telescope costs much less than a conventional telescope of equivalent power, and many more instruments on this pattern are now being planned (for instance, for the observatory on *Mauna Kea).

Multiple star A star made up of more than two physically-associated components. A famous example is *Castor. Another well-known multiple star is Theta Orionis, in the *Orion Nebula.

Mundrabilla Meteorite A 12-ton meteorite found in Western Australia.

Mural quadrant This is described under the heading *Quadrant.

N

N galaxies External *galaxies with bright, condensed nuclei. They are often highly active.

NASA (National Aeronautics and Space Administration) The American agency, set up in 1958, responsible for all non-military space operations by the United States.

NGC (The New General Catalogue of Clusters and Nebulæ) compiled by J. L. E. *Dreyer in the 1880s. It has officially superseded the old *Messier numbers —thus the *Andromeda Spiral, M.31, is also NGC 224—though the M numbers are still widely used.

Nadir The point on the celestial sphere immediately below the observer. Obviously the nadir is directly opposite to the overhead point or *zenith.

Nano- Abbreviation for 10^{-9} or $1/1,000,000,000$ (one thousand millionth). Thus a nanometre is one thousand-millionth of a metre, and a nanosecond is one thousand-millionth of a second.

Nasmyth, James (1808-1890) British

engineer; a pioneer lunar observer, and also the inventor of the steam-hammer.

Neap tide The tide produced when the Sun and Moon are at right angles to the Earth—that is to say, when the Moon is at *quadrature, and its pull is opposed by that of the Sun.

Nebula A mass of 'dust' and tenuous gas in space. If there are suitable stars in or near it, the nebula becomes visible, either by straightforward reflection or because the stellar radiation causes the nebular material to emit a certain amount of light on its own account. If there are no suitable stars, the nebula will be dark, and will be

Spiral nebula in Ursa Major (Mount Wilson and Palomar).

M. 51, the Whirlpool Nebula (Ron Arbour).

detectable only because it blots out the light of objects beyond.

The most famous of the bright nebulæ is M.42, the Sword of Orion, which is easily visible with the naked eye. It contains hot early-type stars (notably the 'Trapezium', Theta Orionis) and is an *emission nebula; it is in fact merely the most obvious part of a vast molecular cloud covering most of Orion. On the other hand, the nebula in the *Pleiades cluster is of the reflection type. Of dark nebulæ, the best-known example is the *Coal Sack in the Southern Cross, but there are many others—for instance, in Cygnus.

Nebulæ are the birthplaces of stars. The process goes on continuously; the dark *globules are probably stars which are not yet hot enough to shine.

Gaseous or 'galactic' nebulæ are also to be found in other galaxies. For example, the *Tarantula Nebula in the Large *Magellanic Cloud dwarfs the Sword of Orion!

Nebular Hypothesis A theory of the origin of the Solar System, put forward in 1796 by the French mathematician *Laplace. According to Laplace, the planets were formed from 'rings' thrown off by a shrinking gas-cloud; the Sun was assumed to be the innermost remnant of the cloud.

The Nebular Hypothesis was widely accepted for many years, but it failed to stand up to mathematical analysis, and had to be rejected. However modern theories follow something of the same principle, inasmuch as it is believed that the planets were formed by accretion from a *solar nebula. See also *Solar System.

Nebular variables Very young stars, not yet old enough to have joined the *Main Sequence and fluctuating irregularly. The prototype star is RW Aurigæ.

Nebularum, Palus (The Marsh of Clouds) Part of the lunar Mare *Imbrium, near the craters Aristillus and Autolycus. The name has been deleted from some maps, but all lunar observers use it.

Nebulium A non-existent element! Unidentified lines in the spectra of gaseous *nebulæ were attributed to such an element, but it is now known that the nebulium lines are due to ionized oxygen and nitrogen.

Nectaris, Mare (The Sea of Nectar) A well-defined lunar 'sea', leading off the Mare *Fœcunditatis. The most prominent crater on it is Rosse. The great bay Fracastorius lies on its southern border, and closely outside it, to the west, are the three large formations *Theophilus, Cyrillus and Catharina.

Negative eyepiece An *eyepiece made up of two lenses, with the image located between them. The Huyghenian eyepiece is of this type.

Neison, Edmund (1851-1940) His real name was Neville. His great book *The Moon* was published in 1876. In 1882 he became the first (and only) Director of the Natal Observatory, in Durban; when he retired in 1912 the Observatory was closed. Neison was also a skilled chemist—and an excellent tennis player!

Neptune The outermost giant planet. For data, see *Planets. It was discovered in 1846 by J. Galle and H. *D'Arrest, at the Berlin Observatory, as a result of calculations by U. J. J. *Le Verrier. Similar calculations had been made independently, in England, by J. C. *Adams.

Neptune is of magnitude 7.7, and therefore well below naked-eye visibility, though binoculars will show it. It is believed that there is a rocky core, with metals and silicates, surrounded by an icy mantle of methane, water and ammonia, which is overlaid by the gaseous atmosphere, made up mainly of hydrogen together with a considerable quantity of helium.

It is often said that Neptune and Uranus are twins, but there are important differences. Neptune is slightly the smaller, but appreciably the more massive; unlike Uranus, it has an internal heat-source. The Neptunian atmosphere is not clear and transparent to great depths, as with Uranus; it contains aerosols—possibly ice crystals—and the 'haze' is variable. Certainly the Uranus/Neptune pair is very different from the Jupiter/Saturn pair.

Whether or not Neptune has a ring we do not yet know. By the 'occultation' method there have been reports of an incomplete ring, but we must presumably await the results from *Voyager 2, due to by-pass Neptune in 1989. The two confirmed satellites are *Triton and *Nereid.

Nereid The second satellite of *Neptune. For data, see *Satellites. It is distinguished by its highly eccentric orbit. It is extremely faint, and hard to see even with giant telescopes.

Nestor Asteroid No 659; one of the *Trojans. It has a mean opposition magnitude of 15.8, and a period of 12.1 years.

Comparative sizes of Neptune and the Earth.

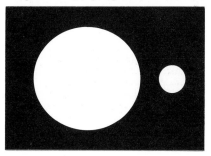

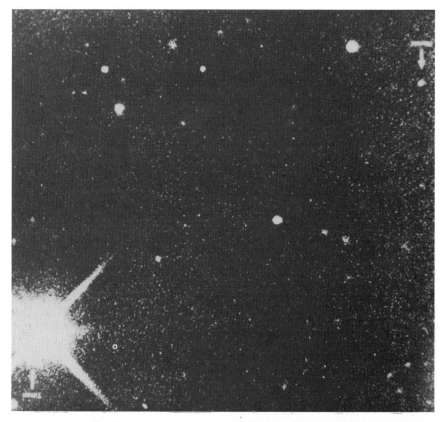

Neptune and satellites.

Neutral point A point on the line joining the centres of mass of two bodies in the position where their gravitational pulls cancel each other out. For the Earth/Moon system, the neutral point is much closer to the Moon, which has only 1/81 of the Earth's mass. In his famous novel *From the Earth to the Moon*, published in 1865, Jules Verne misused the neutral point; he wrote that his space-travellers lost all sensation of weight on reaching it. Actually, the travellers would have been weightless from the moment when they entered *free fall; and in any case, the neutral point is always moving, because of the relative motions of the Earth and Moon. Moreover, the Sun's gravitational pull is always the dominating factor.

Neutrino A fundamental particle with no electrical charge. It has long been assumed that the 'rest mass' is zero, but it has recently been suggested that they may have very slight rest mass, in which case the effects upon cosmological theories would be very marked. Neutrinos are very hard to detect, though some have been trapped by interactions with chlorine, as in the *Homestake Mine experiment.

Neutron star The remnant of a very massive star which has suffered a Type II *supernova outburst. Following the collapse of the star's core, the protons and electrons run together, making up neutrons, and the density is amazing; perhaps 100 million million times that of water. A cubic inch of neutron star material would weigh about 4,000 million tons.

A typical neutron star has a mass about equal to that of the Sun, but may be no more than a few miles in diameter. It seems that there may be an outer layer made up of rigid crystalline iron; inside this is neutron-rich material, and beneath comes a region of superfluid material, largely neutrons but with some protons and electrons. In the centre there is a solid core made up of 'hyperons', which are more fundamental than neutrons, but about which we know practically nothing.

Neutron stars have very strong magnetic fields, and are rotating rapidly. Some at least send out radio waves (see *Pulsars).

Newcomb, Simon (1835-1909) American astronomer; a leading authority on celestial mechanics.

Newton, Sir Isaac (1642-1727) Greatest of all mathematicians. His *Principia*, published in 1687, introduced the 'modern' era of astronomy.

Newton's Laws of Motion These were laid down in the *Principia*. They are:

1. A body continues in its state of rest, or uniform motion, unless acted upon by some external force.

2. If a body is accelerated by an external force F, the acceleration a is directly proportional to the force and inversely proportional to the mass, m, of the body: in other words $F = ma$.

3. Every action has an equal and opposite reaction.

Newtonian reflector The most common type of reflecting telescope. It is named after Sir Isaac *Newton, who built the first telescope of its type around 1671.

With a Newtonian, the light from the object to be studied passes down an open tube on to a curved (parabolic) mirror. This mirror sends the light back up the tube on to a smaller, flat mirror placed at an angle of 45°. The light is then directed into the side of the tube, where an image is formed and is magnified by an *eyepiece. In a Newtonian reflector, therefore, the observer looks into the tube instead of up it. A certain amount of light is lost because the flat gets in the way, but the loss is not serious.

Newtonians are very convenient. The mirrors, of course, need periodical attention; the glass surface is coated with a very thin layer of silver, aluminium or rhodium, which has to be renewed when it becomes tarnished. The telescope tube may be of skeleton construction. In some cases a prism is used in place of a flat secondary.

Nicol prism A prism used in stellar *photometry to reduce the light from the comparison star so that it can be made equal to the variable, or any other star which is to be measured.

Nix Olympica See *Olympus Mons.

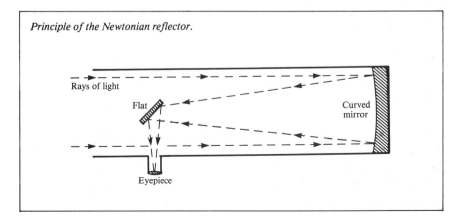

Principle of the Newtonian reflector.

Rays of light

Flat

Curved mirror

Eyepiece

Noctilucent clouds Strange clouds in the ionosphere, best seen at night when they continue to catch the rays of the Sun. They are over 50 miles high, and are quite different from ordinary clouds. It is thought that they are due to dust left by meteors which have entered the upper atmosphere.

Nodes The points at which the orbit of the Moon, planet, comet or asteroid cuts the plane of the *ecliptic. When the body crosses the plane of the ecliptic as it moves from south to north, it is said to pass the *ascending node*; when the movement is from north to south, the body is said to pass the *descending node*. The line joining these two points is called the *line of nodes*.

North America Nebula Nickname for the nebula NGC 7000, near Deneb in Cygnus. It is just visible with binoculars. Photographs show that its outline really does recall the shape of the North American continent.

North polar distance The angular distance of a body from the north celestial *pole.

North Polar Sequence A list of 96 stars, near the north pole of the sky, whose photographic *magnitudes have been measured as accurately as possible. Other stars are then photographed and compared with those in the North Polar Sequence, so that their own magnitudes can be worked out. The Sequence includes stars from magnitude 2 (Polaris) down to as faint as magnitude 20.

Nova A star which suffers a sudden outburst, and flares up to many times its normal brilliancy, remaining bright for a few days, weeks or even months before fading back to obscurity. A few novae have become really spectacular; V.603, Nova Aquilae 1918, reached magnitude −1.1. In some cases the decline is rapid; thus in 1975, V.1500 Cygni reached magnitude 1.8, increasing from near-invisibility to maximum in only a few hours, but it faded to below magnitude 6 within a week, whereas HR Delphini, which flared up in 1967, remained visible with the naked eye for months, and was still above magnitude 12 in 1985.

Novae are not due to collisions between stars. This plausible-sounding theory was abandoned long ago; stellar collisions must be extremely rare even in crowded parts of the Galaxy, whereas novae are not particularly uncommon. It is now thought that a nova is a *binary system, one member of the pair being a normal Main Sequence star while the other component is a White Dwarf. The White Dwarf pulls material away from its companion, and eventually the dwarf is surrounded by a dense layer of gas, largely hydrogen. When enough material has been collected, and the temperature has reached about ten million degrees, thermonuclear reactions are triggered off, and there is a sudden, violent outburst.

A few novae have been observed to suffer more than one outburst; these are *recurrent novae such as the 'Blaze Star', T Coronae. There may be a link between recurrent novae and the so-called 'dwarf novae' or *SS Cygni variables.

The following novae since 1900 have reached magnitude 4 or brighter:

Nova	Year	Maximum Magnitude	Discoverer
GK Persei	1901	0.0	Anderson
DN Geminorum	1912	3.3	Enebo
V.603 Aquilae	1918	−1.1	Bower
V.476 Cygni	1920	2.0	Denning
RR Pictoris	1925	1.1	Watson
DQ Herculis	1934	1.2	Prentice
CP Lacertae	1936	1.9	Gomi
CP Puppis	1942	0.4	Dawson
V.533 Herculis	1963	3.2	Dahlgren and Peltier
HR Delphini	1967	3.7	Alcock
V.1500 Cygni	1975	1.8	Honda

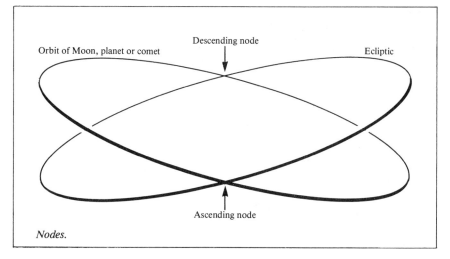

Orbit of Moon, planet or comet

Descending node

Ecliptic

Ascending node

Nodes.

Many novæ have been observed in outer galaxies, but their great distances mean that they are very faint. Note that there is no close connection between novæ and the much more violent *super-novæ.

Nubeculæ Old name for the *Magellanic Clouds.

Nubium, Mare Major lunar 'sea' extending from the *Oceanus Procellarum; the Mare *Humorum, Sinus *Medii and Sinus *Aestuum adjoin it. On its floor are various features, including the Fra Mauro group of ruined ring-plains near which Apollo 14 landed.

Nutation A slight slow 'nodding' of the Earth's axis, due to the fact that the Moon is sometimes above and sometimes below the *ecliptic and therefore does not always pull on the Earth's equatorial bulge in the same direction as the Sun. The result is that the position of the celestial pole seems to 'nod' by about 9 *seconds of arc to either side of its mean position, in a period of 18 years 220 days. The effect is superimposed on the more regular shift of the celestial pole caused by *precession.

Nysa Asteroid No 44. It has a period of 3.8 years, and a diameter of about 50 miles. It has the distinction of being the most

reflective asteroid known, with an *albedo of 0.38.

O

OAO (Orbiting Astronomical Observatories) American artificial satellites. There were four in all, launched between 1966 and 1972, but only two were successful: OAO 2 (1968), which operated until 1973 and sent back important data, including studies of the magnetic fields of stars, ultra-violet emissions from a supernova, and a vast hydrogen cloud around a comet (Tago-Sato-Kosaka); and OAO 3 (the *Copernicus satellite), which went up in 1972 and had a very long active lifetime of over eight years. It carried an 82 cm ultra-violet telescope together with X-ray telescopes. One very important discovery was that there is much more *deuterium ('heavy hydrogen') in inter-stellar clouds than had been expected.

OGO (Orbiting Geophysical Observatories) Six American satellites launched between 1964 and 1969 to study the space environment close to the Earth. Other studies included details of interstellar hydrogen clouds, and the hydrogen envelope of Bennett's Comet of 1970.

OSO (Orbiting Solar Observatories) Nine American satellites, launched between 1962 and 1975; eight were successful, and carried out studies of the Sun at X-ray, ultra-violet and infra-red wavelengths. One important discovery was that of *coronal holes. OSO-8 ceased transmitting in September 1978.

Oberon The fourth major satellite of Uranus. For data, see *Satellites. Little is known about it as yet, but spectra indicate that its surface is icy. It and *Titania are the largest of Uranus' satellites.

Object-glass The main lens of a *refracting telescope. It is also known as an *objective*.

Objective prism A small prism, mounted in front of the main optics of a telescope. The effect is to produce a small-scale spectrum of every star in the field of view, so that many stars can be studied with one photographic exposure. Of course the stellar spectra produced in this way are not detailed, but the method has been found to be very useful.

Oblateness The degree of flattening of a globe. That of the Earth is 0.003 (the equatorial diameter is 7,926 miles, the polar diameter 7,900); that of Saturn is as much as 0.1.

Obliquity of the ecliptic The angle between the *ecliptic and the celestial *equator; it amounts to 23°26′24″. It may also be defined as the angle by which the Earth's axis is tilted from the perpendicular to the orbital plane.

Observatory The usual meaning of the word is 'a building which houses a telescope', but properly speaking an observatory is a full-scale astronomical research station, containing equipment of all sorts.

Early observatories had no telescopes, but were fitted with measuring instruments, such as *quadrants, to measure the apparent positions of the stars and other bodies in the sky. There are various remains to be seen at Delhi and elsewhere; the buildings are remarkably elaborate. The last great observatory of pre-telescopic times was set up by *Tycho Brahe on the Danish island of Hven, and it was here that Tycho used his quadrants to draw up his very accurate star-catalogue. He left Hven for good in 1596, and nothing now remains of his observatory, *Uraniborg.

When telescopes were invented, observatories were naturally developed to house them. Early national observatories were those of Paris and Copenhagen. Greenwich Observatory dates from 1675, and has always been regarded as the time-keeping centre of the world. The first major American observatories were not founded until the second half of the 19th century.

Because of the need for clear skies and steady *'seeing', modern observatories are situated at high altitude whenever possible. The *Pic du Midi Observatory, in the Pyrenees, is about 10,000 ft above sea-level; the observatory on *Mauna Kea almost 14,000 ft. Unfortunately some observatories have been badly affected by light pollution and 'smog'; Mount *Wilson is being virtually closed, and the main instruments at Greenwich were shifted first to *Herstmonceux, in Sussex, and latterly to *La Palma in the Canary Islands.

Today, with the electronic revolution and the extensive use of computers, the astronomer who wants a night's observing need not be in the actual observatory, or even in the same continent; the telescopes can be operated by remote control. A new development will be the setting-up of observatories in space, which will certainly happen in the near future (the *Hubble telescope will be a start).

Amateur astronomers have their own observatories, some of which are highly sophisticated. The forms vary; some are domes, some sheds with sliding roofs, and some are run-off sheds. Everything really depends upon the main interest of the owner.

Above *Colour-enhanced image of Saturn as photographed by Voyager 1 from 21 million miles.*

Below *Clouds on Saturn. Photograph from Voyager 2.*

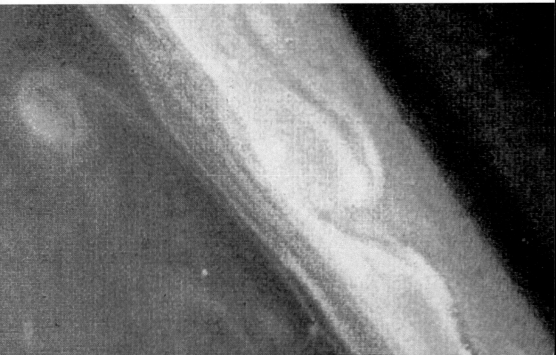

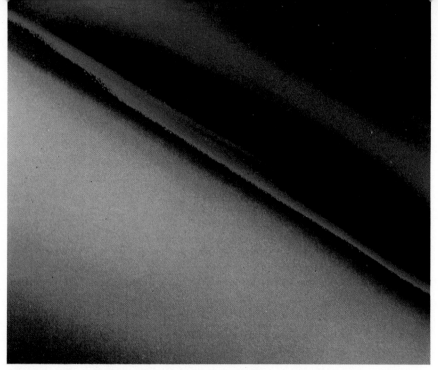

Above *A colour-enhanced picture of the surface of Titan taken by Voyager 1 in 1980 from a distance of 13,700 miles.*

Below *The rings of Saturn photographed from Voyager 1.*

Above *Iapetus photographed by Voyager 2.*

Below *Another view of Saturn's rings seen by Voyager 2.*

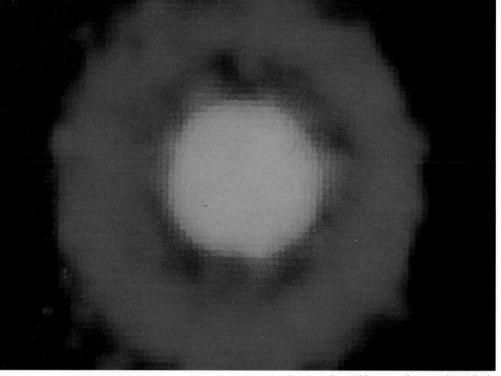

Above *Uranus photographed by infra-red, showing the cold core, the mantle and the comparatively warm atmosphere.*

Below *M.45—the Pleiades, or Seven Sisters.*

Occultation The covering-up of one celestial body by another. Thus the Moon may pass in front of a star, or (occasionally) a planet; a planet may occult a star; and there may have been cases when one planet has occulted another—for instance, Venus occulted Mars in 1590. Strictly speaking, a solar *eclipse is merely the conventional name for the occultation of the Sun by the Moon.

Occultations of stars by the Moon have always been regarded as important. Were the Moon's edge surrounded by an appreciable atmosphere, the star would flicker and fade briefly before disappearing, but in fact it does not; it snaps out abruptly. One moment it is visible; the next, it has been blotted out by the advancing Moon. The apparent positions of the stars are known very precisely, but until modern times the exact position of the Moon was not so well determined. Therefore, if an occultation were timed, the position of the Moon's limb at that moment was also known.

Radio sources can also be occulted; in fact the position of the first *quasar was determined by this method.

In recent years, occultations of stars by small bodies such as asteroids and planetary satellites has provided a means of measuring the apparent diameters of the occulting bodies. It was studies of this kind which led to the discovery of Pluto's satellite, *Charon.

Ocular An alternative name for *eyepiece.

Caldera of Olympus Mons, Mars.

Odessa Crater An impact crater in Texas, discovered in 1921. It has a diameter of 558 ft.

Odin Planitia A plain on Mercury, adjoining the *Caloris Basin.

Olbers, Heinrich Wilhelm Matthäus (1758-1840) Skilful German amateur astronomer, who concentrated upon asteroids—he discovered *Pallas and *Vesta—and upon comet orbits. By profession he was an eminent medical doctor. He was one of the members of the *Celestial Police.

Olbers' Comet A periodical comet with a period of 69.5 years; it has now been seen at three returns, the last being that of 1956, but it is never a conspicuous object.

Olbers' Paradox A paradox discussed by *Olbers in 1826: Why is it dark at night? If the universe is infinite, then sooner or later each line of sight will encounter a star! It is now explained as a result of the expansion of the universe, since the light from very remote objects will be *redshifted into invisibility. Moreover the observable universe, at least, is not infinite.

Olympus Mons The highest Martian volcano, rising to a height of 15 miles above the adjacent surface. It is topped by a complex 49-mile caldera, and is associated with what appears to be an extensive drainage system. It adjoins the *Tharsis Ridge. It is clearly visible from Earth, with a powerful telescope, as a

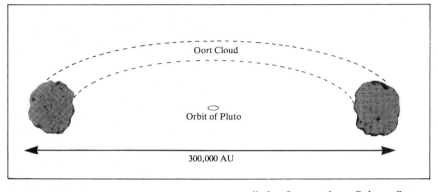

Oort Cloud (not to scale).

minute patch; it was formerly known as Nix Olympica—the Olympic Snow.

Omega Nebula Messier 17, a gaseous *emission nebula in Sagittarius. It is almost 5,900 light-years away, and since it is of the 7th magnitude it is an easy telescopic object; it lies close to the star Gamma Scuti. It is also a radio source. An alternative name for it is the Horseshoe Nebula.

Oort Cloud According to a theory by the Dutch astronomer J. H. Oort, there is a 'cloud' of comets orbiting the Sun at a distance of between 30,000 and 100,000 *astronomical units. When perturbed for any reason, a comet swings inward toward the Sun. If it is 'captured' by a planet, it becomes a short-period comet; if not, it either returns to the Oort Cloud or else is put into an open orbit and expelled from the Solar System altogether.

Ophiuchid Meteors A minor shower, reaching maximum about 20 June. The usual *ZHR is about 10.

Open cluster A loose or galactic *cluster.

Opposition The position of a planet when it is exactly opposite to the Sun in the sky. At opposition, the planet, the Sun and the Earth lie in approximately a straight line, with the Earth in the mid position. Obviously, the *inferior planets, Mercury and Venus, can never come to opposition.

Optical double A double star in which the two components are not genuinely associated, but merely happen to lie in almost the same line of sight as seen from Earth.

Optical window The region of the *electromagnetic spectrum in which

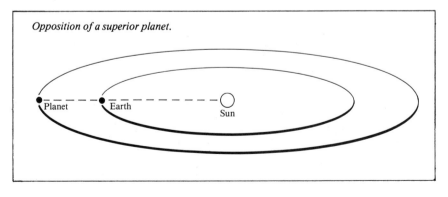

Opposition of a superior planet.

radiations can pass through the atmosphere and reach the surface of the Earth. It extends from about 300 to 900 *nanometres (3,000 to 30,000 *Ångströms).

Orbit The path of a celestial object.

Orbiters Five unmanned American space-craft which were put into closed orbits round the Moon and provided maps of virtually the whole of the lunar surface. All were successful; Orbiter 1 was launched in August 1966; Orbiter 5, last of the series, in August 1967. The programme ended with the final controlled impact of Orbiter 5 on the Moon's surface on 31 January 1968.

Orientale, Mare (The Eastern Sea) An important lunar mare—a vast ringed structure on the extreme limb of the Moon as seen from Earth, extending on to the far side. It was discovered and named by H. P. Wilkins and myself; at that time the 'east' limb of the Moon was taken to be in the opposite sense to that now adopted, so that as seen from the Earth our *Eastern* Sea is now on the *western* limb of the Moon!

Orion Arm A spiral arm of the *Galaxy; the Sun lies close to its edge.

Orion Nebula The bright gaseous *emission nebula in the Sword of Orion; its Messier number is M.42. Photographs show that M.42 and the less conspicuous nebula M.43 are concentrations of much more extensive nebulosity, and in fact it is now known that they are merely the visible parts of a vast molecular cloud covering most of Orion.

The Orion Nebula is distinctly visible with the naked eye, and moderate telescopes show it well. Immersed in the nebulosity is the *multiple star Theta Orionis, known as the Trapezium for reasons which will be obvious to anyone who has seen it.

The distance of the Orion Nebula is about 1,200 light-years; its density is amazingly low, no more than one-millionth of the best vacuum which we can produce in our laboratories. It is a stellar birthplace; fresh stars are being formed

The Orion Nebula.

from the nebular material and there are numbers of very young *T Tauri variables. Deep in the Nebula is the infra-red *Becklin-Neugebauer Object, which may be a very luminous star, and also the *Kleinmann-Low Object. The total diameter of the visible nebula is about 25 light-years.

Orionids A meteor shower reaching its maximum about 21 October; the usual *ZHR is about 30. The Orionids are associated with *Halley's Comet.

Orrery A model showing the *Solar System, with the planets capable of being moved at their correct relative velocities round the Sun by some mechanical device (such as turning a handle). The name comes from Charles Boyle, the fourth Earl of Orrery, who lived during the early 18th century.

Oscillating universe The theory according to which the universe alternately expands and contracts, so that there are '*Big Bangs' at intervals of perhaps 80,000 million years.

Outgassing The processes by which gases escape from the crust of a planet into its atmosphere, by vulcanism.

Owl Nebula Messier 97, a *planetary nebula in Ursa Major. It is about 10,000 light-years away, and as its integrated magnitude is about 12 it is by no means a conspicuous telescopic object. The two stars seen in it do give an impression of an owl's face, when photographed with a large instrument.

Ozma In 1960 radio astronomers at Green Bank in West Virginia, under the direction of F. Drake used the 85-ft radio telescope in a serious search for intelligent life beyond the Solar System. They studied one particular wavelength—21 centimetres—which is the wavelength of the radiation emitted by clouds of cold hydrogen in the Galaxy—in the hope that they might pick up some signal-pattern which could be shown to be artificial. The experiment was not successful, and was discontinued after a few months, but it has been succeeded by other programmes of the same type. The original experiment was officially called Project Ozma, after the 'Wizard of Oz' in the famous children's story by Frank Baum, though it was more popularly known as Project Little Green Men!

Ozonosphere A layer in the Earth's atmosphere, between 7½ and 30 miles above sea-level, in which there is a relatively high concentration of ozone (O_3). The ozonosphere blocks out radiations from space which would be harmful to life, so that without it we could not survive.

P

P Cygni A remarkable variable star in Cygnus. It was recorded as a third-magnitude star in 1600, and then slowly faded until it had dropped below naked-eye visibility. In 1654 it reappeared, and finally settled down to around magnitude 5 by 1715; since then it has shown only minor fluctuations. It is very luminous, and is over 4,500 light-years away.

Pallas The second *minor planet to be discovered. For data, see *Minor Planets. Unlike *Ceres, it is not of the carbonaceous type; it is classified as 'peculiar carbon'.

Palomar Observatory Major observatory in California; the main instruments are the 200-in *Hale reflector, completed in 1948, and a 48-in *Schmidt camera. For a time Palomar and Mount Wilson were administered jointly, as the Hale Observatories, but this has now been discontinued.

Palomar Sky Atlas Photographic sky atlas, from the north pole to declination 33°S. It was published in the 1950s, and extends down to magnitude 21. The photographs were taken with the 48-in Schmidt telescope.

Panspermia theory A theory proposed by the Swedish chemist Svante Arrhenius in 1906, according to which life on Earth was brought here from outer space by way of a meteorite. The theory was never popular, as it seemed to raise more problems than it solved, though it has recently been revived, in a different form, by *Hoyle and Wickramasinghe.

Parallax, Trigonometrical The apparent angular shift of a distant body when observed from two different directions.

The best way to show what is meant is to make a practical experiment. Shut one eye, and line up your finger with an object some way away, such as a clock on the mantelpiece. Now, without moving your finger or your head, use the other eye.

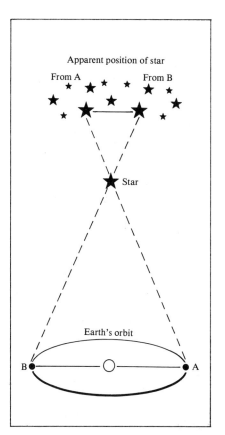

Apparent position of star

From A   From B

★ Star

Earth's orbit

B ●   ○   ● A

Your finger will no longer be aligned with the clock. If you know the distance between your eyes, and also the angular shift of your finger against the background, it is mathematically possible to work out the distance between your finger and your eyes. Half the angle of displacement is known as the trigonometrical parallax.

In 1838 the German astronomer F. W. Bessel applied this method to a star, 61 Cygni, which he believed might be relatively close (because it has considerable *proper motion, and is also a wide *binary). He measured the position of the star twice, at an interval of six months. He was therefore observing the star from opposite sides of the Earth's orbit; and since the Earth-Sun distance is 93 million miles, he was using a baseline of twice this length, or 186 million miles. He found that 61 Cygni showed a measurable parallax against the background of more remote stars, and he was able to show that the distance was about eleven light-years. Many other stars have since had their distances measured in the same way, but beyond 300 light-years or so the parallax

Left *Parallax. The near star will be displaced as seen from A and B, opposite ends of the Earth's orbit.*

Below *Dome of the 200-in reflector at Palomar.*

shifts become so small that they cannot be measured accurately.

Parkes Observatory Leading Australian radio astronomy observatory, in New South Wales; the main instrument is the 210-ft 'dish'. There is very close collaboration between Parkes and the optical observatory at *Siding Spring.

Parking orbit A closed orbit round the Earth in which a space-craft is put before being sent into its final planned orbit. Thus the *Giotto probe to *Halley's Comet made three revolutions in a parking orbit before being boosted into its trajectory toward the comet.

Parsec The distance at which a star would show a *parallax of one second of arc. It is equal to 3.26 *light-years, 206,265 *astronomical units, or 19,150,000 million miles. Actually, no star apart from the Sun is as close as this; the nearest star, Proxima Centauri, is over 4 light-years from us (1.3 parsecs) and has a parallax of 0".79.

Parsons, William See *Rosse, Lord.

Pasiphaë The eighth satellite of Jupiter. For data, see *Satellites. It was lost after its discovery in 1908, found again in 1922, lost once more until 1938 and again between 1941 and 1955!

Pavonis Mons A lofty Martian volcano on the *Tharsis Ridge. The summit caldera is 28 miles in diameter.

Payload The mass which is put into orbit by rocket power—in other words, the total mass of the vehicle minus that of the launcher or launchers.

Peculiar stars Stars whose spectra cannot be fitted neatly into any class; denoted by p. Thus the spectrum of *P Cygni is B1p.

Pele An Ionian volcano. At the time of the pass of *Voyager 1 it was the most active volcano on the satellite. By the time of the *Voyager 2 pass it was no longer erupting, but is unlikely to have become extinct.

Penumbra The area of partial shadow to either side of the main cone of shadow cast by the Earth. Its effects are described under the heading *Eclipses, Lunar.

The term is also applied to the outer, relatively light parts of sunspots.

Perigee This is described together with *Perihelion.

Perihelion The position in the orbit of a planet or other body when at its nearest to the Sun. For instance, the Earth is at perihelion in early January, when the distance between the two bodies is 91½ million miles; at *aphelion, in early July,

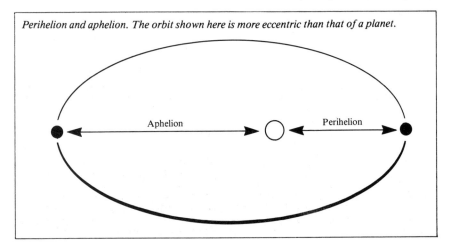

Perihelion and aphelion. The orbit shown here is more eccentric than that of a planet.

Aphelion   Perihelion

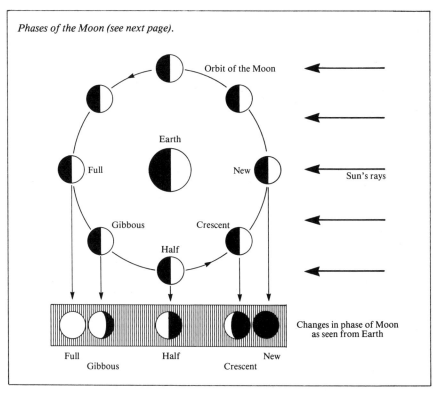

Phases of the Moon (see next page).

Orbit of the Moon

Earth

Full

New

Sun's rays

Gibbous

Crescent

Half

Changes in phase of Moon
as seen from Earth

Full   Half   New
Gibbous   Crescent

the distance has increased to 94½ million miles. Similarly, perigee refers to a body moving round the Earth.

Periodic time See *Sidereal Period.

Period-luminosity law The relationship between the period and the real luminosity of a *Cepheid or *RR Lyræ variable.

Perrine, Charles Dillon (1867-1951) Discoverer of eleven comets and two of the satellites of Jupiter (Himalia and Elara). He was for some years Director of the Argentine National Observatory at Cordoba.

Perrine Regio One of the major regions on *Ganymede.

Perseids The main annual meteor shower. For data, see *Meteors. The parent comet, *Swift-Tuttle, has not been seen since 1862.

Perseus Arm A spiral arm of the *Galaxy, about 7,000 light-years from the Sun.

Perturbations The disturbances in the orbit of a celestial body produced by the gravitational pulls of others. For instance, *Neptune was tracked down because of its perturbations on *Uranus. A body of slight mass, such as a *comet, may have its orbit violently perturbed if it passes relatively close to a massive body such as a planet, and the comet's period may be completely altered—or even changed into an open curve.

Petavius A major lunar crater, 106 miles in diameter. It lies in the south-west quadrant, and is one of a chain of great walled formations (the others are Furnerius, Vendelinus and Langrenus). Petavius has high walls, a central mountain group, and a prominent *rill crossing the floor.

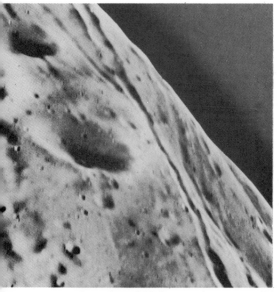

The surface of Phobos seen from a mere 120 km by Viking Orbiter 1 in 1977. The area shown is about 3 by 3.5 km (NASA).

Phæthon The asteroid which makes the closest known approach to the Sun. It is only a mile or so in diameter.

Phase angle The angle between the Earth and the Sun, as seen from another body or from space.

Phases The apparent changes of shape of a celestial body from new to full. The Moon has no light of its own, and depends entirely upon light reflected from the Sun. When the Moon is almost between the Sun and the Earth, its dark side is turned toward us, and it cannot be seen (unless the alignment is perfect, resulting in a solar *eclipse). When the Moon is on the far side of the Earth with respect to the Sun, its lighted half is turned toward us, and the Moon is full; at other times it may be crescent, half or *gibbous.

Mercury and Venus also show complete phases from new to full. Mars may be decidedly gibbous except when near opposition, but the other planets are so much further away that their phases are inappreciable.

Phillips, Theodore E. R. (1868-1942) English clergyman and amateur astronomer, noted for his studies of Jupiter.

Phobos The inner and larger satellite of Mars; for data, see *Satellites. Phobos has a cratered and grooved surface; the largest crater, Stickney, is 3 miles in diameter. Phobos was the first known natural satellite to have a revolution period shorter than the rotation period of its primary. Its mean opposition magnitude is above 12, but its closeness to Mars makes it a difficult object to observe except with a large telescope.

Phocylides, Johannes Phocylides Holwarda (1618-1651) Dutch astronomer, who was one of the first to assume that the stars have *proper motions; in 1638 he rediscovered *Mira Ceti, and was the first to realize that it is a variable star. A lunar walled plain in the south-west quadrant of the Moon is named after him.

Phœbe The ninth satellite of Saturn. For data, see *Satellites. It has retrograde motion, and may be a captured asteroid. It does not have *captured rotation; its revolution period is 550 days, its rotation period only 9 hours. It is less well-known than the other satellites, since neither *Voyager probe passed close to it, but it seems to have a darkish surface, and could be of the same type as the *carbonaceous asteroids.

Phœnicids A minor meteor shower, reaching its maximum on 4 December. The average *ZHR is about 5. Since the radiant lies in Phœnix, it is well seen only from southern latitudes.

Phosphorus The old name for *Venus as an evening object.

Photoelectric cell An electronic device. Light falls upon the cell, and produces an electric current; the strength of the current depends upon the intensity of the light.

Photoelectric photometer An instrument used for measuring the magnitudes of stars or other objects. It consists basically of a *photoelectric cell used with the telescope, and can be made remarkably sensitive.

Photography, Astronomical The word 'photography' was coined by an astronomer, Sir John *Herschel, in the earlier half of the last century. For more than a hundred years now it has been possible to take good astronomical photographs, and by about 1900 photography had superseded visual observation for almost all branches of astronomy, partly because the human eye is easily deceived and partly because it is obviously easier to study pictures in the comfort of a laboratory than to make drawings and notes at the eye-end of a telescope. Moreover, the sensitive plate is more efficient than the eye, and can provide photographs of objects which are beyond visual range. Only now is photography giving way to electronic devices. It must however be stressed that photography is still of tremendous value scientifically, as has been shown by the recent results obtained by David Malin at the *Siding Spring Observatory in Australia.

Ordinary astronomical photography can be carried out by amateurs, and spectacular pictures of the Moon and stellar objects can be taken, provided that the telescope used with the camera is firmly mounted and well guided.

Photometer An instrument used to measure the intensity of light coming from one particular source. There are various types. The old *wedge photometer* used a sliding scale, darkened to different degrees along its length; the star was observed at various points, and the position in which the star became invisible was noted. However, modern photometers are always *photoelectric.

Photometry The measurement of the intensity of light.

Photomultiplier A complicated form of *photoelectric cell, in which the original current produced is amplified many times.

Photon The smallest 'unit' of light.

Photosphere The bright surface of the *Sun.

Piazzi, Giuseppe (1746-1826) Italian astronomer, and Director of the Palermo Observatory in Sicily. In 1801 he discovered the first minor planet, *Ceres.

Pic du Midi Observatory French observatory in the Pyrénées, at an altitude of 9,400 ft. It is noted particularly for the excellence of its planetary photographs, taken with the 24-in refractor.

Pickering, E. C. (1846-1919) Pioneer American stellar spectroscopist; for many years Director of the Harvard College Observatory. He made many contributions, including the detection of the first *spectroscopic binary (Mizar A).

Pickering, W. H. (1858-1938) Brother of E. C. Pickering. His main work was in connection with the Solar System; he discovered Saturn's satellite *Phœbe (the first satellite to be found by photographic methods) and produced one of the first good photographic atlases of the Moon, showing each section of the lunar surface under several different conditions of lighting and *libration.

Pico A prominent lunar mountain, south of *Plato. It is triple-peaked, and rises to about 8,000 ft.

Pioneer probes Unmanned American space-craft. Data are as follows:

Number	Launch year	Target	Comments
1	1958	Moon	Failure.
2	1958	Moon	Failure.

continued overleaf

Number	Launch year	Target	Comments
3	1958	Moon	Failure, though discovered the Earth's second radiation belt.
4	1959	Moon	By-passed Moon, entered solar orbit.
5	1960	solar orbit	Returned solar *flare and *wind data.
6	1965	solar orbit	Returned solar data.
7	1966	solar orbit	Solar data; Earth's *magnetosphere.
8	1967	solar orbit	Data on solar wind; cosmic rays; solar flares; the magnetosphere.
9	1968	solar orbit	Solar data.
10	1972	Jupiter	Successful. Now leaving Solar System.
11	1973	Jupiter	Successful. Also by-passed Saturn. Now leaving Solar System.
12	1978	Venus	Orbiter and entry probes. Successful.

Pioneers 10 and 11 carry plaques, which will (it is hoped!) enable any subsequent 'finders' to identify their system of origin! Contact should be maintained with them until the 1990s, and it is even possible that they may provide information leading to the tracking-down of a planet beyond Neptune and Pluto, though this is admittedly very much of a 'long shot'.

Piton A lunar mountain on the Mare *Imbrium, not far from *Pico. It is bright, and rises to 7,000 ft.

Plages, Solar Bright regions on the Sun's surface, observed in the light of one element only (hydrogen or calcium). Also termed *flocculi.

Planetarium An instrument used to show an 'artificial sky' on the inside of a large dome, and able to reproduce celestial phenomena of all kinds. The first modern-type projector was installed at Jena, East Germany, in 1923; it was manufactured by the firm of Carl Zeiss, and designed by Professor Walther Bauersfeld. Most major cities now have planetaria.

Planetary Nebula The name is misleading, since a planetary nebula is not a true nebula and is certainly not a planet. It is in fact an old star which has thrown off its outer layers, so that it is surrounded by a 'shell' of expanding gas. The central stars are small and very hot, with surface temperatures of the order of 100,000°C. There may be up to 50,000 planetary nebulæ in our Galaxy alone. Obviously a planetary is short-lived in this form, and after about 50,000 years the gas will have expanded so far that it has become very rarefied, and ceases to shine. The most famous planetary nebula is M.57, the *Ring Nebula in Lyra; another is the Dumbbell Nebula, M.27, in Vulpecula, which is less symmetrical.

Planetesimals Small bodies which grow into planets, by the process of accretion.

Planetoid An alternative name for a *minor planet or asteroid. It seems to have become virtually obsolete now.

Planets The most important members of the Sun's family:

Planet	Mean distance from the Sun in miles	Sidereal Period	Synodic Period in days	Orbital Eccentricity	Orbital Inclination ° '
Mercury	36,000,000	88 days	115.9	0.206	7 0
Venus	67,200,000	224.7 days	583.9	0.007	3 24
Earth	92,957,000	365.3 days	—	0.017	—
Mars	141,500,000	687 days	779.9	0.093	1 51
Jupiter	483,300,000	11.9 years	398.9	0.048	1 18
Saturn	886,100,000	29.5 years	378.1	0.056	2 29
Uranus	1,783,000,000	84.0 years	369.7	0.047	0 46
Neptune	2,793,000,000	164.8 years	367.5	0.009	1 46
Pluto	3,667,000,000	247.7 years	366.7	0.248	17 10

Planet	Mean orbital velocity in miles/second	Axial rotation	Axial inclination	Oblateness	No of Satellites
Mercury	29.8	58.65 days	Low	Negligible	0
Venus	21.8	243.16 days	178°	0	0
Earth	18.5	23 hours 56 minutes 4 seconds	23°27'	0.003	1
Mars	15.0	24 hours 37 minutes 23 seconds	23°59'	0.009	2
Jupiter	8.1	9 hours 50 minutes 30 seconds	3°5'	0.06	16
Saturn	6.0	10 hours 39 minutes	26°44'	0.1	20
Uranus	4.2	± 17 hours	98°	0.06	5
Neptune	3.4	± 17 hours 57 minutes	28°48'	0.02	2
Pluto	2.9	6 days 9 hours 17 minutes	?	Low	1

Planet	Diameter miles (equatorial)	Mass Earth = 1	Volume Earth = 1	Density water = 1	Surface gravity Earth = 1
Mercury	3,033	0.055	0.056	5.5	0.38
Venus	7,523	0.815	0.86	5.25	0.90
Earth	7,926	1	1	5.52	1
Mars	4,218	0.107	0.15	3.94	0.38
Jupiter	88,378	318	131	1.33	2.64
Saturn	74,145	95	744	0.71	1.16
Uranus	32,190	14.6	67	1.7	1.17
Neptune	30,760	17.2	57	1.8	1.2
Pluto	1,800	Very low	—	Very low	Below 0.1

Planet	Escape velocity miles/second	Albedo	Mean surface temperature °C	Maximum Magnitude	Mean apparent diameter of Sun, seen from planet
Mercury	2.6	0.06	350 (day), − 170 (night)	− 1.9	1°22'40"
Venus	6.4	0.76	480	− 4.4	44'15"
Earth	6.94	0.36	22	—	31'59"
Mars	3.2	0.16	− 23	− 2.8	21'
Jupiter	37.1	0.43	− 150	− 2.6	6'09"

Planet	Escape velocity miles/second	Albedo	Mean surface temperature °C	Maximum Magnitude	Mean apparent diameter of Sun, seen from planet
Saturn	22.0	0.61	− 180	− 0.3	3′22″
Uranus	13.9	0.35	− 210	5.6	1′41″
Neptune	15.1	0.35	− 220	7.7	1′04″
Pluto	Very low	0.4?	− 230	14	0′49″

Pluto has an eccentric orbit, and at present is closer-in than Neptune. Its planetary status is in doubt (see *Pluto). Some of the data given here, such as the albedo values, are uncertain to some extent. The naked-eye planets were of course known in ancient times; Uranus was discovered in 1781, Neptune in 1846 and Pluto in 1930. There is a very strong probability that at least one still more distant planet exists.

Planets, Extra-solar It is highly probable that other stars have planetary systems of their own. No telescope yet made will show a planet of another star directly, but various nearby stars show irregularities in their *proper motions which may be due to planetary companions; for instance *Barnard's Star, a red dwarf only 6 light-years away, may have at least two planets. However, the main evidence comes from infra-red observations. In 1983 the *IRAS satellite detected large infra-red excesses from about forty stars, including *Vega, *Fomalhaut and *Beta Pictoris, which could be due to cool, planet-forming

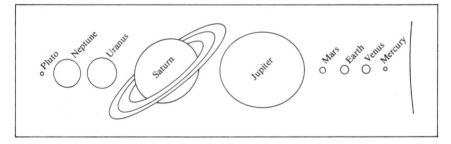

Above *Comparative sizes of the planets.*

Below *Orbits of the planets to scale.*

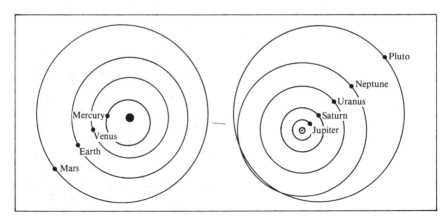

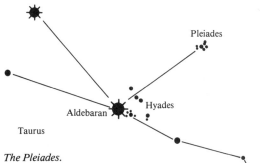

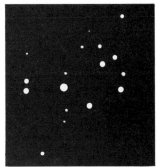

The Pleiades.

material, and with Beta Pictoris this has now been recorded visually.

Plaskett, John Stanley (1865-1941) Canadian astronomer, noted for his researches into the nature of hot, early-type stars. He was the first director of the Dominion Astrophysical Observatory in Victoria, BC, and supervised the construction of the 82-in mirror for the McDonald Observatory in Texas.

Plaskett's Star HD 47129 Monocerotis, identified by J. S. *Plaskett in 1922. It is a binary; each component is believed to have about 55 times the mass of the Sun, making it the most massive stellar system known.

Plasma Gas in which the atoms are wholly ionized. A plasma is therefore made up of positively or negatively charged particles (protons, atomic nuclei or electrons) moving independently.

Plato A 60-mile lunar crater, named after the great Greek philosopher (BC 427-347). The crater is noted for its very dark floor, which makes it readily identifiable whenever it is in sunlight. It lies between the Mare *Frigoris and the Mare *Imbrium.

Pleiades The most famous of all open *clusters, usually known as the Seven Sisters. The brightest star is Alcyone or Eta Tauri, magnitude 2.9. Next come Electra, Atlas, Merope, Maia, Taygete, Celæno, Pleione and Asterope. Observers with normal sight can see at least seven stars under average conditions; the record is said to be nineteen. Many more stars are shown telescopically. There are several hundred stars in the cluster, together with a beautiful reflection nebula; the distance is 410 light-years. The cluster is No 45 in *Messier's list.

Pleione is a remarkable 'shell star', somewhat variable in brightness. All the leading stars of the cluster are hot and white, and their age has been estimated at less than a hundred million years.

Plerion A *supernova remnant which is brightest at its centre, such as the *Crab Nebula.

Plough The nickname for the pattern made up of the seven chief stars of Ursa Major, the Great Bear: Alkaid, Mizar, Alioth, Megrez, Phad (or Phekda), Merak and Dubhe. Five of these make up a *moving cluster; Alkaid and Dubhe

The Pleiades.

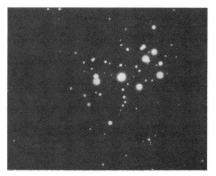

153

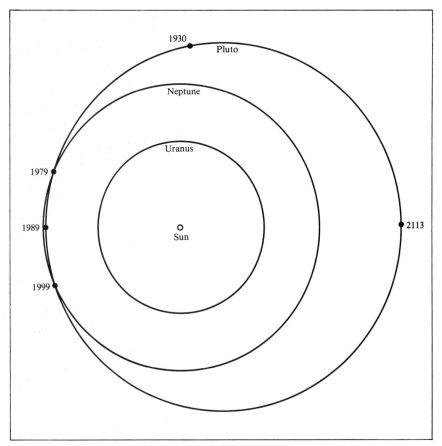

Above *Orbit of Pluto.*

Below *Orbits of the outer planets (not to scale).*

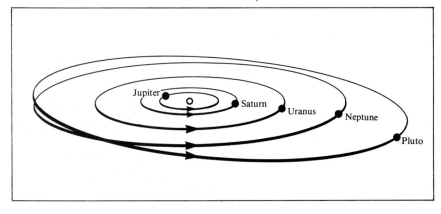

have *proper motions in the opposite direction, so that eventually the Plough will lose its familiar shape.

Pluto The ninth planet, discovered by Clyde *Tombaugh in 1930 as a result of a systematic search from the *Lowell Observatory in Arizona. Its orbit is much more eccentric than that of any other planet. For most of its 248-year period it is much further out than Neptune, but near perihelion it is closer-in; however, the orbit is inclined at the relatively sharp angle of 17°, so that there is no chance of a collision. The next perihelion passage is due in 1989, so that between 1979 and 1999 Pluto temporarily forfeits its title of 'the outermost planet'.

Pluto has been found to be very small, with a diameter less than that of the Moon, so that it could not cause any measurable *perturbations in the movements of giant worlds such as Uranus and Neptune—which means that its discovery was largely fortuitous, since its position had been calculated, by Percival *Lowell, on the basis of just these perturbations! (It must be added, however, that Tombaugh's search was extensive, and not confined to the area given by Lowell.)

Pluto is of low density, and is probably made up of a mixture of rock and ice, with a surface layer of methane ice. In 1978 it was found to have a companion, *Charon, which has a diameter of about one-third that of Pluto itself. Its revolution period is equal to Pluto's axial rotation period (6 days 9 hours), which had already been determined by variations in the apparent magnitude. This is a situation unique in the Solar System. It seems doubtful whether Pluto is worthy of true planetary status; it may be an exceptional asteroid (or, more properly, asteroid pair). As the magnitude never rises much above 14, it is beyond the range of small telescopes.

Pogson's Ratio The ratio between the brightness of two stars of successive magnitude. It is logarithmic; the ratio is 2.512, so that a star of magnitude 1 is 100 times as bright as a star of magnitude 6 (since 2.512 is the fifth root of 100). Variable star observers train themselves

Clyde Tombaugh, discoverer of Pluto when I photographed him at the Lowell Observatory in 1982 with the blink-microscope which he used in the search for Pluto.

to estimate differences of 0.1 magnitude; this is often known as Pogson's Step Method.

Polaris The star Alpha Ursæ Minoris, which lies within one degree of the north celestial pole. For data see *Stars.

Polarization Normal light-waves vibrate in all planes. If light is polarized, the light vibrates in one plane only. Studies of the polarization of light are of great importance; for instance, the light from remote objects is polarized when it passes through clouds of interstellar dust, because of the orientation of the dust particles in the presence of a magnetic field.

Poles, Celestial The north and south points of the *celestial sphere. The north pole is marked within a degree by *Polaris; there is no bright south polar star, the nearest naked-eye star to the polar point being the obscure *Sigma Octantis.

Pollux The star Beta Geminorum, the brighter of the two Twins (the other is *Castor). For data, see *Stars. In ancient times Pollux was rated below Castor in brightness, but it is now much the more brilliant of the two. On the whole, however, it is not likely that there has been any real change, and it is more probable that we are dealing with an error in interpretation or translation. Pollux has a K-type spectrum, and is very obviously orange.

Pond, John (1767-1836) The sixth Astronomer Royal (1811-1835). His early administration was fruitful, and he was an excellent observer; but ill-health later handicapped him, and he was asked to resign office. He was succeeded by *Airy.

Pons, Jean Louis (1761-1831) French astronomer, who began his career as a doorkeeper at the Marseilles Observatory and ended it as Director of the Museum Observatory in Florence. He specialized in comet-hunting, and has a record 36 discoveries to his credit.

Pons-Brooks Comet A comet with a period of 71 years, discovered by *Pons in 1812 and recovered by W. R. Brooks in 1883; the latest return was that of 1954. The comet reaches the fringe of naked-eye visibility, and may develop a considerable tail.

Pons-Winnecke Comet A comet with a period of 6.3 years, discovered by *Pons in 1819 and recovered by F. A. Winnecke in 1858; since 1869 it has been missed at only three returns. In 1827 it reached magnitude 3.5, but since then the orbit has been perturbed by Jupiter, and the comet is usually faint (below magnitude 11).

Populations, Stellar Two main types of star regions. Population I consists of hot white stars, together with a great deal of interstellar material in the form of dust and gas; the brightest stars of Population II are old red giants, and there is relatively little interstellar material, indicating a much greater age. No hard and fast boundaries can be laid down, but the arms of spiral galaxies are mainly of Population I, while the central regions and the halo areas, as well as elliptical galaxies and *globular clusters, are mainly of Population II.

Position angle The apparent direction of one object with reference to another, measured from the north point of the main object through east (90°), south (180°) and west (270°) back to north (360° or 000°). Position angles of double stars are measured by means of *micrometers.

Poynting-Robertson effect The spiralling down of small particles into the Sun, because of their interactions with solar radiation.

Præsepe Messier 44, an open cluster in Cancer, easily visible with the naked eye. It is 520 light-years away. Unlike the *Pleiades cluster, it contains no detectable nebulosity, and is presumably much older (perhaps 400 million years); many of

Position angle.

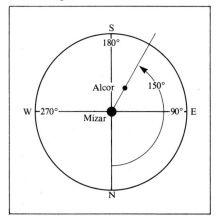

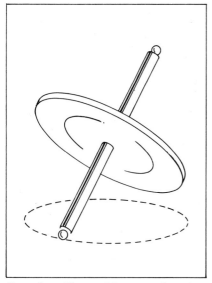

Precession, illustrated by a toppling gyroscope.

slow motion has taken the First Point out of Aries into the neighbouring constellation of Pisces, though the name has not been changed. In ancient times, too, the north celestial pole lay close to the star *Thuban or Alpha Draconis. At the moment Polaris occupies the position of honour, but by AD 12,000 we will have a much more brilliant pole star, *Vega.

Precession of the equinoxes has been known for a long time. It was discovered by the Greek astronomer *Hipparchus in the second century BC.

Prime meridian The meridian on the Earth's surface which passes through both poles and also the Airy *transit instrument at the Old Royal Observatory in Greenwich Park. It is taken as longitude 0°, and marks the boundary between the eastern and the western hemispheres. The decision to adopt the Greenwich meridian as the zero for longitudes was taken in the late 19th century by international agreement.

Prism A triangular or wedge-shaped block of glass. Light passing through it is split up, because the different wavelengths are bent or *refracted* by different amounts—the shorter the wavelength, the greater the amount of refraction. Prisms are essential parts of most instruments based upon the principle of the *spectroscope, though in some cases *diffraction gratings are preferred.

In a *Newtonian reflector, a special prism may replace the more conventional flat mirror used to send the light from the main speculum into the side of the tube. If this is done, there is no splitting-up of the light into a spectrum, since parts of the prism are silvered, though on the whole a flat mirror is far better than a prism for this purpose.

Procellarum, Oceanus (The Ocean of Storms) The largest of all the lunar 'seas', though much less regular than some others. It leads into the Mare *Imbrium, Mare *Nubium and Sinus *Roris.

Procyon Alpha Canis Minoris; for data see *Stars. Procyon has a faint White Dwarf companion, discovered in 1896 by

its leading stars are of late spectral type, and so are yellow or orange. Præsepe is often nicknamed the 'Beehive'. It is easy to locate, near the two naked-eye stars Delta and Gamma Cancri.

Precession The apparent slow movement of the celestial poles.

The Earth is not a perfect globe; its equator bulges slightly, and the Moon pulls upon this bulge. The result is that the Earth's axis seems to describe a small circle in the manner of a gyroscope which is starting to topple. The circle on the *celestial sphere is only 47° in diameter, and takes 25,800 years to complete, but the effects are important. Because the poles shift, the celestial *equator also moves. This in turn shifts the position of the vernal *equinox or *First Point of Aries, which is where the equator cuts the *ecliptic. Since the *right ascensions and *declinations of stars are measured from the First Point of Aries and the celestial equator, these, too, will alter slightly from year to year.

The First Point of Aries moves by 50.4 *seconds of arc yearly along the ecliptic, from west to east. Since ancient times, this

157

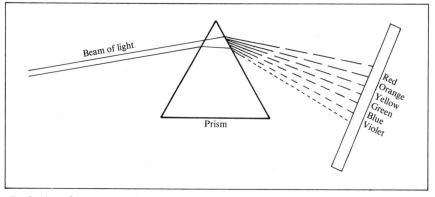

Production of a spectrum using a prism.

J. M. Schaeberle with the 36-in *Lick refractor. The period of revolution is 40.7 years, and the mean separation about 1,400 million miles.

Prometheus A very active volcano on *Io, which was erupting at each of the *Voyager encounters.

Prominences, Solar Masses of glowing hydrogen above the Sun's surface. Many are associated with sunspots, and they are of immense size; the length of an average prominence is over 100,000 miles. They are of two main types. *Eruptive prominences* are in violent motion, while *quiescent prominences* are relatively calm, so that they may hang in the Sun's atmosphere for several weeks.

Prominences can be seen with the naked eye only when the Sun's surface is hidden by the Moon during a total *eclipse, but filters can show them at any time. They were formerly known as Red Flames, but this misleading name has long since been dropped.

Proper motion The individual movement of a star on the *celestial sphere. All stellar proper motions are very slight, because the stars are so remote; the *constellation patterns remain to all intents and purposes unaltered over periods of many consecutive human lifetimes, though over a sufficiently long period the patterns will alter. The star with the largest proper motion is *Barnard's Star: 1 minute of arc every 6 years. It therefore takes 180 years to move by an amount equal to the apparent diameter of the full moon. The

Proper motions of stars in the Great Bear.

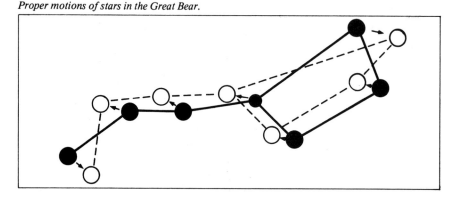

proper motions of distant stars in our Galaxy, and of other galaxies, are too slight to be measurable at all.

Proton A fundamental particle with a positive electrical charge. The nucleus of the simplest atom, that of hydrogen, consists of a single proton; the nuclei of other atoms are made up of protons and *neutrons. Since a neutron has no electrical charge, it follows that the total charge of an atomic nucleus is positive. This positive charge is balanced out by the combined negative charges of the *electrons moving round the nucleus, so that the complete atom is electrically neutral.

Proton-proton reaction One of the ways in which a star may produce its energy. As with the *carbon-nitrogen cycle, the final result is that helium is formed from hydrogen, with loss of mass and release of energy. It is now known that the proton-proton reaction plays the main role in stars such as the Sun.

Proxima Centauri The nearest star beyond the Sun, at a distance of 4.2 light-years. It is a companion of the bright binary *Alpha Centauri; it was discovered by R. T. Innes, in 1913. Proxima is a dim red dwarf, of type M5, with only 1/13,000 the luminosity of the Sun; if it were to

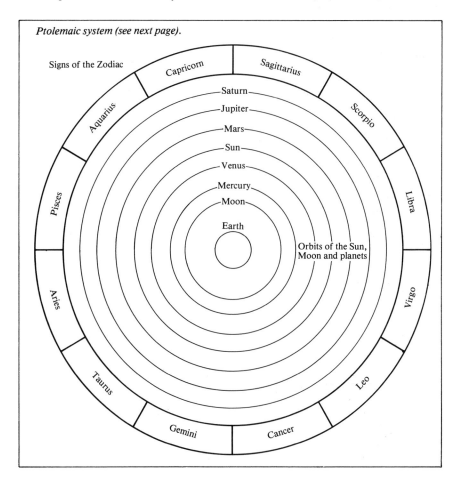

Ptolemaic system (see next page).

replace the Sun it would send us only about as much light as 45 full moons. The diameter is probably about 40,000 miles, smaller than Jupiter or Saturn. Proxima is also a *flare star.

Ptolemæus, Claudius (Ptolemy) (*c* AD 120-180) Last great astronomer of Classical times. Much of our knowledge about ancient science comes from his book, which has come down to us via its Arab translation as the *Almagest. Occasional attempts to discredit Ptolemy have been notably unsuccessful.

Ptolemæus A large lunar walled plain, over 90 miles in diameter, near the centre of the visible disk; it is one of a chain of three major formations, the others being Alphonsus and Arzachel. Ptolemæus has low walls, and a relatively flat floor containing no really large craterlets.

Ptolemaic system The old plan of the universe, according to which the Earth lay motionless in the centre with all other bodies revolving round it. Ptolemy did not invent it, but he brought it to its highest degree of perfection, and certainly it fitted the facts as he knew them. Not until more than a thousand years after Ptolemy's time was it seriously challenged.

Domes at the Pulkova Observatory, Leningrad.

Pulkova Observatory Major Russian observatory, near Leningrad. It was founded in 1839, but completely destroyed by the Germans during the war; reconstruction was completed in 1954. There are several large telescopes, and also a radio telescope. Optically, observing conditions are poor; during summer the sky never becomes really dark, while in the winter there is a great deal of cloud—which is why the newer Russian observatories have been set up in the southern part of the USSR.

Pulsar A rapidly-varying radio source, now known to be a *neutron star. The first pulsar was discovered by Jocelyn Bell-Burnell, at Cambridge, in 1967 while a completely different investigation was being carried out. The source (in Vulpecula) pulsed so rapidly and so regularly that for a brief period it was even thought that the signals might be artificial! When this intriguing idea was discarded, it was thought that the object might be a rapidly-spinning White Dwarf, but this theory too had to be abandoned—small though it may be, a White Dwarf could not possibly vibrate so quickly. The neutron star theory is now unchallenged. Few pulsars have been identified with optical objects; the first was that in the *Crab Nebula, and the second the pulsar in *Vela. Pulsars seem to be slowing down as they lose energy; this is true even of the Crab, which has a current period of 0.033

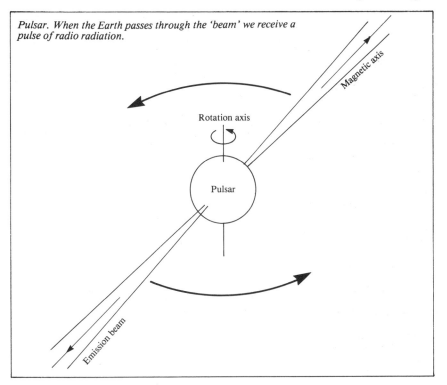

Pulsar. When the Earth passes through the 'beam' we receive a pulse of radio radiation.

Magnetic axis

Rotation axis

Pulsar

Emission beam

second. It follows that pulsars must be comparatively short-lived, in this stage of their evolution. Recently some ultra-fast pulsars have been identified which are not slowing down, and which seem to be of rather different type.

Pulsating variable A variable star in which the changes are intrinsic, with alternate swelling and shrinking of the star. *Cepheids, for example, are of this type.

Purkinje effect The effect of colour upon estimations of the brightness of a light source. If a red star is to be compared with a bluish one, the Purkinje effect may cause serious errors in deciding which really is the brighter.

Putredinis, Palus The Marsh of Decay; part of the Mare *Imbrium, between the Apennines and the craters of the Archimedes group.

Pythagoras (*c* BC 562-500) Great Greek geometer, who was one of the first, if not the very first, to maintain that the Earth is not flat. He also seems to have studied the movements of the planets. A 70-mile lunar crater near the Moon's north-western limb, with high, terraced walls and a massive central elevation, has been named after him.

Q

Q Cygni A nova discovered by J. Schmidt in 1876. It reached the third magnitude, but has now become very faint.

QSO Quasi-stellar object; see *Quasar.

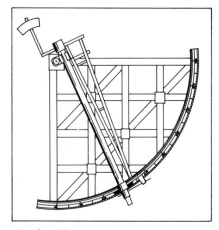

Mural quadrant.

Quadrant An astronomical measuring instrument used in former times. It consisted of an arc graduated into 90°, together with a sighting pointer; it was used to measure the apparent positions of celestial bodies. Since quadrants were often large, they were attached to walls, and were then known as mural quadrants (Latin *murus*, a wall).

Quadrantids A meteor shower, with a short, sharp maximum on 3 January. At times the Quadrantids may be spectacular, and the *ZHR can reach 110. There is no known parent comet. The shower is named because the radiant lies in the area covered by the now-rejected constellation of Quadrans, the Quadrant; it is now included in Boötes.

Quadrature When the Moon or a planet is at right angles to the Sun as seen from Earth, it is said to be at quadrature. Thus the Moon is at quadrature when it is half-phase.

Quantum The energy possessed by one *photon of light. A quantum is therefore the smallest amount of light-energy which can be transmitted at any given wavelength.

Quark A very fundamental particle, from which many elementary particles are formed. None has yet been actually observed. It has been suggested that the inner regions of *neutron stars may be made up largely of quarks.

Quartz clock A clock which is regulated by the vibrations of a quartz crystal.

Quasar A very remote, incredibly luminous object. The term 'quasar' has now generally superseded the original 'quasi-stellar object' or QSO.

In 1963 Maarten Schmidt, at Palomar, examined the spectrum of an object which seemed to coincide in position with a strong radio source. The object looked like a bluish star when seen through a telescope, but the spectrum showed it to be something quite different. There was no similarity to the spectrum of a normal star, and the lines showed tremendous *red shifts, which presumably meant that the object was receding at high velocity. This, in turn, meant that it had to be extremely luminous—at least 200 times brighter than a whole galaxy.

The first quasar caused intense interest, and a search was made for other objects of the same sort. Many were found, not all of which were powerful at radio wavelengths. For some years they remained a complete mystery, but it now seems that they are the nuclei of very active galaxies, though it is still not known whether all galaxies go through a 'quasar period' at some stage in their evolution. The most distant quasar so far identified, PKS 2000-330, is thought to be 13,000 million light-years away, and to be receding at more than 90 per cent of the velocity of light. It is widely believed that quasars are powered by massive central *Black Holes. All are very remote. It must be added that a few astronomers maintain that the red shifts in quasar spectra are not Doppler effects, and that the quasars are not nearly so distant as is usually believed; among these dissentients are H. C. Arp and Sir Fred *Hoyle. It is, however, very much of a minority view.

Quasars had been photographed often enough in the past, but not until 1963 was it realized that they were anything but faint stars in our own Galaxy. Their identification was one of the most

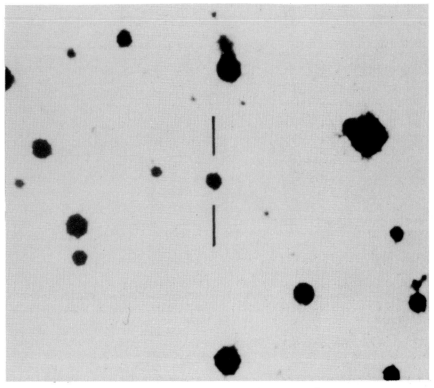

The most distant quasar, PKS 2000-330.

important developments of the second half of our own century.

Quételet, Jacques (1796-1874) Belgian astronomer, best remembered for his work on *meteor radiants; it was he who first identified the *Perseid shower.

Quetzalcoatl An *Amor-type minor planet, discovered in 1953; its asteroid number is 1915. In its discovery year, 1953, it passed within 5 million miles of the Earth, and it is one of the few asteroids to have been detected by radar (others are *Eros, Psyche, Klotho, *Apollo, *Icarus and *Toro). The diameter of Quetzalcoatl is no more than two miles; the period is 4.5 years.

Quiet Sun The state of the Sun when near the minimum of its eleven-year cycle of activity, as in the mid-1980s.

R

R Coronæ Borealis A variable star in the Northern Crown which has given its name to a whole class of such stars. For most of time the brightness remains at maximum, with only minor fluctuations, but there are sudden, unpredictable falls to minimum followed by a slower rise back to normal brightness. R Coronæ itself is usually about magnitude 6, on the fringe of naked-eye visibility, but at minimum it may fall to as low as magnitude 15. Stars of this type are uncommon. They are

163

usually F or K type supergiants; they are poor in hydrogen but rich in carbon, and it is thought that minima occur when carbon particles (soot) accumulated in the star's atmosphere block out part of the light until they are blown away.

RR Lyræ variables Formerly termed cluster-Cepheids, because many of them are found in star clusters (though the prototype, RR Lyræ, is not a cluster member). They have amplitudes of no more than 2 magnitudes, and periods of from 0.05 to 1.2 days. They appear to be about equal in luminosity (around 90 times as powerful as the Sun) and can therefore be used as 'standard candles' in the same way as *Cepheids. Most of them are giants of types A to F.

RV Tauri variables Pulsating variables, mainly G to K type giants. There are alternative deep and shallow minima, though at times the variations become completely irregular. The brightest member of the class is R Scuti, which can reach magnitude 5.7.

RW Aurigæ variables Irregular variables, with rapid and unpredictable fluctuations. Like the *T Tauri stars, they are very young by stellar standards.

Radar astronomy A branch of astronomical science which dates only from the end of the war. It depends on the principle of radar: a pulse of energy is transmitted, and 'bounced back' from a distant object, so that the 'echo' is received and measured. The only everyday comparison is to picture what happens when a tennis-ball is thrown against a wall and caught on the rebound. The analogy is by no means accurate, but it serves to give the general idea.

Radar echoes have been obtained from many bodies in the Solar System, out to and including Saturn. Since a radar pulse moves at the same speed as light, the time-delay between the transmission of the pulse and the reception of the echo gives the distance of the target—a method which has been used to give a very precise measurement of the distance of Venus, and hence, from *Kepler's Laws, of the

*astronomical unit. Radar equipment carried in space-probes has also enabled very correct maps of Venus to be obtained. Meteors can also be studied by radar, since for this purpose a meteor trail behaves in much the same way that a solid body would do. By now, radar observations of meteors have largely replaced the old visual methods.

Radial velocity The towards-or-away movement of a celestial body: positive if the object is receding, negative if it is approaching. All galaxies except those of the *Local Group show positive radial velocities.

Radiant The point in the sky from which the *meteors of any particular shower seem to radiate. For instance, the August shower has its radiant in Perseus—hence the name *'Perseids'.

Radio astronomy A branch of astronomy which began less than sixty years ago, but has now become of the utmost importance.

In 1931 Karl Jansky, an American radio engineer, was using an improvised aerial to study 'static' when he found that he was picking up radio waves from the sky. They proved to come from the *Milky Way, in the direction of the star-clouds in Sagittarius now known to indicate the direction of the centre of the Galaxy. Jansky's work was followed up by an American amateur, Grote Reber, who built the first real radio telescope in 1939, but only after the war did radio astronomy assume real importance.

Radio waves are electromagnetic vibrations, but they do not affect our eyes, and have to be collected and studied by instruments which are really in the nature of large aerials. After 1947, many discrete radio sources were found. Some, such as the *Crab Nebula, proved to be supernova remnants; the Sun was (predictably) a radio source, and later emissions were detected from Jupiter. However, most of the sources were very remote, lying well beyond our Galaxy. Some external galaxies are surprisingly powerful at radio wavelengths.

All sorts of developments have been

made possible by these methods. Without radio astronomy it is unlikely that either *pulsars or *quasars would have been identified. We have also cleared up one vital point about our own Galaxy. Between the stars there are clouds of cold hydrogen, quite invisible optically, but which emit radio waves at 21.1 cm; the distribution of these clouds has given final conclusive proof that the Galaxy is spiral in form. It is also worth noting that radio waves can be received across immense distances, so that in this way we can reach out to the remote parts of the observable universe.

Radio galaxy A galaxy which is particularly powerful in the radio range. Many *Seyfert galaxies are of this type.

Radio star Obsolete term for a discrete radio source, dropped when it became clear that most sources are not individual stars. However, there are some stars which send out detectable radio waves—*Antares has a companion of this type—and in such a sense the term can be retained.

Radio storm Burst of radio emission from the Sun, usually associated with a solar *flare.

Radio telescope An instrument used for collecting and analyzing radio waves from space. The name is not an apt one, since a radio telescope does not produce a visible picture, and one cannot look through it! The radio waves are collected and focused, and the result usually appears as a trace on a graph.

It is also possible to convert the energy into sound—hence the hackneyed term *radio noise*—but the actual sound is produced inside the equipment. Where there is no air to carry sound-waves, there can be no noise, and there is no air in space.

Some radio telescopes are 'dishes' or *paraboloids*, such as the 250-ft steerable dish at *Jodrell Bank in Cheshire; the 1,000-ft dish at *Arecibo in Puerto Rico is built in a natural hollow in the ground. Other instruments depend upon the principle of interference, and are known as *radio interferometers*. In fact, radio telescopes take many forms, each of which is suited to some particular branch of research.

Radio window The range of radio wavelengths to which the Earth's atmosphere is transparent. It extends from twenty metres down to a few millimetres.

Radiometer An instrument for measuring the total amount of energy emitted by a body in the form of radiation.

Radius vector An imaginary line joining the centre of a planet (or comet) to the centre of the Sun. According to *Kepler's Laws, the radius vector sweeps out equal areas in equal times.

Ramsden eyepiece An *eyepiece in which two identical plano-convex lenses are used, with their convex faces together. The *Kellner* is a form of Ramsden, with an achromatic eye-lens.

Ranger probes Unmanned American lunar vehicles, crash-landed on the Moon between 1961 and 1965; their role was to obtain close-range pictures before impact. The last three only were successful. Ranger 7 (1964) landed in the Mare Nubium, Ranger 8 (1965) in the Mare Tranquillitatis, and Ranger 9 (1965) in the crater Alphonsus. The results were much the best obtained up to that time.

Ras Algethi The star Alpha Herculis. It is a huge red supergiant, over 200 light-years away and on average 700 times as luminous as the Sun; it is variable between magnitudes 3.0 and 3.8. The spectrum is of type M. It is a semi-regular variable, with a very rough period of about 100 days; the diameter is at least 300 million miles, larger than that of the orbit of Mars. There is a 5.4-magnitude greenish companion at a distance of 4.6 seconds of arc.

Ras Alhague The star Alpha Ophiuchi. It is of magnitude 2.08; type A5; distance 62 light-years; luminosity sixty times that of the Sun.

Rayet, Charles Antoine (1839-1906)

165

French astronomer, who became Professor of Astronomy at Bordeaux. Together with Charles Wolf, he drew attention to the very hot stars now known as *Wolf-Rayet stars.

Rays, Lunar Bright streaks associated with some lunar craters, extending outward in all directions; the main systems are those of *Tycho and Copernicus, while others are associated with Kepler, Anaxagoras and Olbers. They are surface features only well seen under high solar illumination; near full moon the Tycho rays dominate the entire lunar scene.

Recombination When a free *electron is captured by an ion, the ionization energy and kinetic energy of the electron are radiated, producing a 'recombination glow'.

Recurrent nova A star which has been known to show more than one nova-like outburst. The best-known example is the 'Blaze Star', *T Coronæ, which flared up to naked-eye visibility in 1866 and again in 1946.

Red shift When a luminous body is receding, the *Doppler effect means that its spectral lines will be shifted over to the long-wave or red end of the spectrum. The greater the shift, the greater the velocity of recession. The red shifts in the spectra of all galaxies beyond the *Local Group show that the whole universe is in a state of expansion.

Reflection grating A device consisting of a large number of parallel rulings on a reflecting surface; it is used to obtain high-dispersion spectra.

Reflection nebula A cloud of dust and gas which is lit up by a nearby star, but which—unlike an *emission nebula— does not emit any light of its own. There is a beautiful reflection nebula in the *Pleiades.

Reflector An optical telescope in which the light from the object under study is collected by a curved mirror. Details are given under the headings *Cassegrain, *Gregorian, *Herschelian, *Newtonian, *Maksutov and *Schmidt telescopes.

Refraction The 'bending' or change of direction of a ray of light when passing through a transparent substance. For instance, light is refracted when it passes through a glass lens or prism. The Earth's atmosphere also causes refraction; an object close to the horizon will be seen higher up than it really is, and the effect may amount to more than half a degree so that the Sun or the full moon may sometimes be visible when it is theoretically below the horizon. When on the point of setting, the lower limb of the Sun will be more affected than the upper limb, so that the Sun's disk will appear flattened.

Refractor A telescope in which the light is collected by means of a lens. The light passes through the main lens or *object-glass, and is brought to focus, where the image is magnified by an *eyepiece.

Astronomical refractors have compound object-glasses, as otherwise an object such as a star would show false colour. (In fact, the false colour nuisance can never be completely cured.) A refractor is more effective than a *reflector of the same aperture, but is much more expensive.

The first astronomical refractors were made during the early 17th century. The largest refractor in use today is the 40-in at the *Yerkes Observatory in America; it is unlikely that a larger refractor will be built, partly because it is much easier to make a reflector with superior light-grasp, and partly because a gaint lens tends to distort under its own weight,

Distortion of the setting sun.

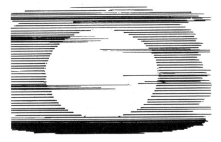

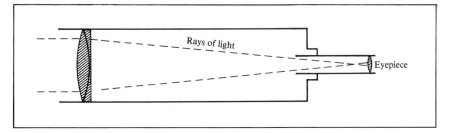

Principle of the refractor.

making it useless. This trouble does not apply to a reflector, since a mirror can be supported at its back, whereas the object-glass of a refractor has to be supported round its edge.

Regiomontanus, Johann Müller (1436-1476) German astronomer, who published printed planetary tables which were the best of their time.

Regolith The outermost 'loose' layer of the Moon or a planetary body. The lunar regolith is a *breccia containing many different ingredients. The average depth is around 15 ft in the maria, over 30 ft in the highlands.

Regression of the nodes The *nodes of the Moon's orbit move slowly westward, making one full circuit of the orbit in 186 years. This regression is due to the gravitational pull of the Sun.

Regulus The star Alpha Leonis, sometimes called 'the Royal Star'. It is the brightest star in the pattern known as the Sickle of Leo. For data, see *Stars.

Relativity theory Developed by Albert *Einstein. The Special Theory was published in 1905 and the General Theory in 1915. Rather than replacing *Newtonian theory, it is better to say that relativity extends it. So far all tests made have supported it.

Resolving power The ability of a telescope to separate objects which are close together. The larger the telescope, the better the resolving power. A good guide is to say that the resolving power R of a telescope of aperture D is $R = 12/D$, where R is given in seconds of arc and D in centimetres.

Rest mass The mass of a particle when motionless. According to *relativity theory, the mass increases with increase of velocity, reaching infinity at the speed of light— which is another way of saying that for a material body, travel at the speed of light is impossible.

Retardation The difference in the time of moonrise from one night to the next. It may exceed one hour, but at the time of *Harvest Moon it may be reduced to as little as a quarter of an hour.

Reticle A series of fine threads placed in the focal plane of an *eyepiece, used for measuring purposes.

Retrograde motion In the Solar System, a body which moves round the Sun in a direction opposite to that of the Earth is said to have retrograde motion. Many retrograde comets are known (including *Halley's) but no retrograde planet or asteroid has been found. The term is also applied to the satellites of planets; the only large satellite of this type is *Triton, the senior satellite of Neptune.

Another meaning of the term refers to the apparent movements of the planets. Usually a planet moves eastwards against the starry background, but we are observing from the Earth, which is itself in motion, so that at various times the planets seem to move in a westward or retrograde direction. Of the planets themselves, Venus has retrograde rotation; so, technically, has Uranus, where

Rhea, the Saturnian moon, photographed from Voyager 1 in November 1980 (JPL).

the axial inclination is 98° to the perpendicular to the orbit.

Reversing layer The gaseous layer above the bright surface or *photosphere of the Sun. Shining on its own, the gas would produce bright spectral lines; but, as the photosphere makes up the background, the lines are reversed, and appear dark. (See also *Spectroscope.) Strictly speaking, the whole of the Sun's *chromosphere is a reversing layer.

Rhæticus, Georg Joachim (1514-1576)

German astronomer, who was largely responsible for persuading *Copernicus to publish his book claiming that the Sun, not the Earth, is the centre of the planetary system.

Rhea The fifth satellite of Saturn. For data, see *Satellites. Rhea is heavily cratered, with an icy surface; as with *Dione, the leading hemisphere is brighter than the trailing hemisphere. It was well surveyed from the *Voyager probes. Rhea is of intermediate size, about equal to *Iapetus and much larger than the *Tethys-Dione pair. As its mean

Culmination.

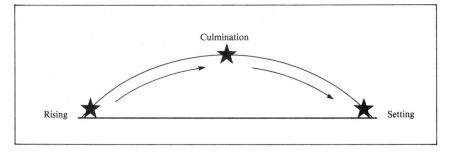

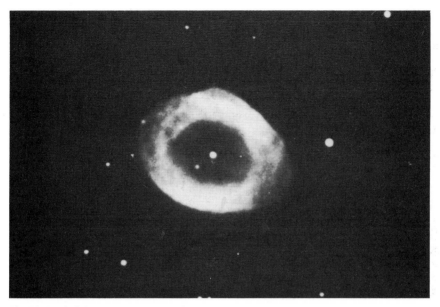

The Ring Nebula in Lyra (see next page).

opposition magnitude is above 10, it is an easy telescopic object.

Rheita Valley A formation in the south-east quadrant of the Moon. It is not a true valley, but is made up of craters which have run together. The Reichenbach Valley, in the same quadrant, is also a crater-chain.

Riccioli, Giovanni (1598-1671) Italian Jesuit; a pioneer of telescopic observation, though he never accepted the theory that the Earth moves round the Sun. In 1651 he published an important lunar map, based on observations by his pupil *Grimaldi, and allotted names to the main craters which are still in use. He named a very large walled plain near the western limb, with a dark floor, after himself. Riccioli also discovered the first telescopic double star, *Mizar.

Rigel Beta Orionis; a particularly luminous B-type star. For data, see *Stars. It has a 7th-magnitude companion; separation 9.2 seconds of arc.

Right ascension The angular distance of a star from the *First Point of Aries or vernal *equinox, measured westward. In practice it is usually given in hours, minutes and seconds of time. The First Point of Aries must reach its highest point (*culmination) once in 24 hours; the right ascension of the star, or other body, is the time-difference between the culmination of the First Point of Aries, and the culmination of the object. For instance, *Rigel in Orion culminates 5 hours 14 minutes 32 seconds after the First Point; the right ascension of Rigel is therefore 5 hours 14 minutes 32 seconds. The usual abbreviation is RA or α.

The right ascensions of bodies in the Solar System change rapidly, but those of stars are constant—apart from the small, slow, regular changes due to *precession.

Rill Crack-like feature on the surface of the Moon. Rills are also known as *rilles* or *clefts*. True rills are collapse features, but many so-called rills are really crater-chains.

Rima An alternative name for a *rill.

169

Ring micrometer A form of *micrometer in which no moving wires are used, as in the *filar micrometer. The ring micrometer is the simpler of the two types, but it is not so accurate.

Ring Nebula Messier 57, a planetary nebula in Lyra between the naked-eye stars Beta and Gamma Lyræ. It is about 1,400 light-years away, and is visible in a small telescope. It is more or less symmetrical, with a faint central star. It is expanding at the rate of about one second of arc per century.

Ring-plains Alternative name for large lunar walled plains or craters.

Ritchey-Chrétien telescope A variant of the *Cassegrain system, giving a sharp image over a wide field of view.

Roche limit The distance from the centre of a planet within which a second body would be broken up by the planet's gravitational pull. This applies only to an orbiting body which has no appreciable 'gravitational cohesion', so that strong objects such as artificial satellites can move safely inside the Roche limit. The Roche limit lies at 2.44 times the radius of the planet from the centre of the globe, so that for the Earth the limit is about 5,700 miles above ground level.

Rocket astronomy A rocket functions by the *principle of reaction*. Its propellants produce hot gases, which are sent out of the vehicle's exhaust, so propelling the vehicle itself in the opposite direction; the rocket 'pushes against itself', so to speak, and will therefore be at its best in space where there is no atmosphere to set up resistance. In 1926 R. H. Goddard, in America, fired the first liquid-propellant rocket; during the war, the Germans developed rockets as military weapons (the *V2 rockets were used to bombard England); but it was not until 1957 that the first *artificial satellite was launched by means of rocket power. Today, of course, rockets have become both powerful and reliable. They have sent men to the Moon, and probes out to many of the planets.

Rømer, Ole (1644-1710) Danish astronomer. He was the first to measure the velocity of light, in 1675 (by observations of the phenomena of Jupiter's satellites), and he also invented various astronomical instruments, notably the transit circle.

Röntgen rays Old name for *X-rays.

Roris, Sinus The Bay of Heats. Lunar bay, connecting the Mare *Frigoris with the *Oceanus Procellarum.

Rosette Nebula NGC 2237-2244 Monocerotis; a beautiful *emission nebula surrounding an open *cluster. It is detectable with binoculars, and when photographed with large telescopes it is a glorious sight. It contains both bright and dark material and many *globules which may be embryo stars.

Rosse, 3rd Earl of (1800-1867) Irish amateur astronomer, who in 1845 completed what was then much the largest reflector in the world. It had a 72-in mirror, and was set up at Lord Rosse's estate at Birr Castle in Central Ireland. Its most famous achievement was the detection of the spiral forms of galaxies.

Rosse, 4th Earl of (1840-1908) He continued his father's work at Birr, and was also the first to make an accurate measurement of the tiny quantity of heat sent to us by the Moon. After his death the 72-in reflector was dismantled, but at the present time it is in the process of being brought back into working order.

Rotation variable A star which shows periodical changes in magnitude because it is rotating, and some parts of its surface are hotter and more luminous than others.

Rover, Lunar (LRV or Lunar Roving Vehicle) 'Moon car' taken to the Moon by Apollo astronauts. There were three in all—those of Apollos 15, 16 and 17. They were left behind on the Moon, and eventually, no doubt, will be recovered.

Royal Observatory, Edinburgh Major British observatory at Blackford Hill,

Edinburgh. It was founded in 1818, and is an important centre for research; for example it has a *Starlink installation. The present Director is Professor Malcolm Longair, who is also Astronomer Royal for Scotland.

Rudolphine Tables Improved tables of planetary motion, published by *Kepler in 1631 and named in honour of the Holy Roman Emperor, Rudolph II.

Runaway Star Nickname for *Barnard's Star, which has the greatest *proper motion known.

Russell, Henry Norris (1877-1957) Leading American astronomer, remembered mainly for his work on stellar evolution and for his part in drawing up the *Hertzsprung-Russell Diagram.

Rutherfurd, Lewis (1816-1892) American spectroscopist; a pioneer of lunar and planetary photography.

Ryle, Sir Martin (1918-1984) Pioneer British radio astronomer, associated mainly with Cambridge; he was knighted in 1966 and became Astronomer Royal in 1972. He was responsible for many fundamental advances, particularly in the design of radio telescopes. It was his observations which disproved the *steady-state theory of the universe.

S

S Andromedæ A *supernova seen in 1885 in the *Andromeda Spiral. It reached the 6th magnitude, but at the time its true nature was not appreciated, and it soon faded into invisibility. It is much the brightest extragalactic supernova ever recorded.

SAO Catalogue An important star catalogue published by the Smithsonian Astrophysical Observatory in 1966. It includes 250,000 stars, and extends down to magnitude 9.

SS Cygni variables Also known as U Geminorum variables or as dwarf novæ. They show frequent outbursts, rising by several magnitudes before returning to their normal minimum brightness. It is now known that SS Cygni stars are *binaries; one component is a red Main Sequence dwarf, and the other is a White Dwarf. The White Dwarf pulls material away from its companion, forming a ring; the addition of further hydrogen-rich material creates an unstable situation, and an outburst occurs.

SS Cygni itself is the brightest member of the class. Its usual magnitude is below 12, but at maxima it can rise to 8.3. The mean period between outbursts is about fifty days.

SS 433 A remarkable stellar system, associated with the radio source W.50. It is an X-ray emitter, and its spectrum is variable. Apparently it is a binary; one member of the pair is ejecting streamers of gas in diametrically opposite directions. The jets are precessing in the manner of a rotating lawn-sprinkler, with the star orbiting its companion in a period of twelve days. It lies in Aquila, and is a faint optical object. Probably it is a supernova remnant of the type now becoming known as a *scintar.

Sabæus Sinus A dark marking on Mars, connecting the *Syrtis Major with the Margaritifer Sinus.

Saiph The star Kappa Orionis; magnitude 2.06, distance over 2,000 light-years, luminosity about 50,000 times that of the Sun. Its spectral type is B. It has been calculated that a million years ago Saiph shone as the brightest star in the sky, with a magnitude of −4.3, equal to that of Venus! It would then have been in the northern hemisphere of the sky.

Salpeter Process ('Triple-alpha Process') A reaction in which three alpha-particles (helium nuclei) form into a carbon nucleus, with the release of energy. A temperature of about a hundred million degrees Centigrade is needed.

Salyut Manned Russian space-stations.

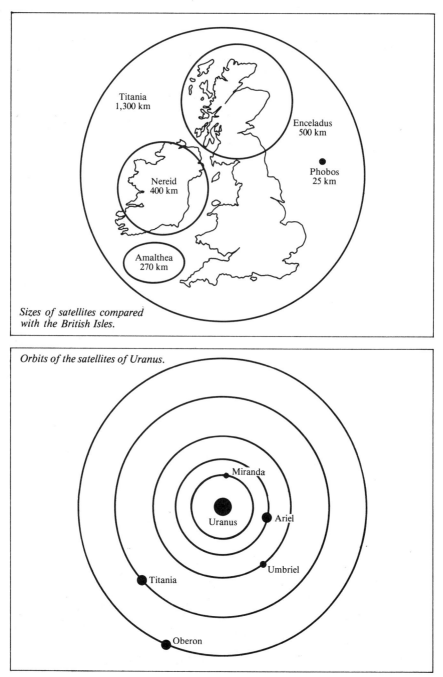

Sizes of satellites compared with the British Isles.

Titania 1,300 km

Enceladus 500 km

Nereid 400 km

Phobos 25 km

Amalthea 270 km

Orbits of the satellites of Uranus.

Miranda

Ariel

Uranus

Umbriel

Titania

Oberon

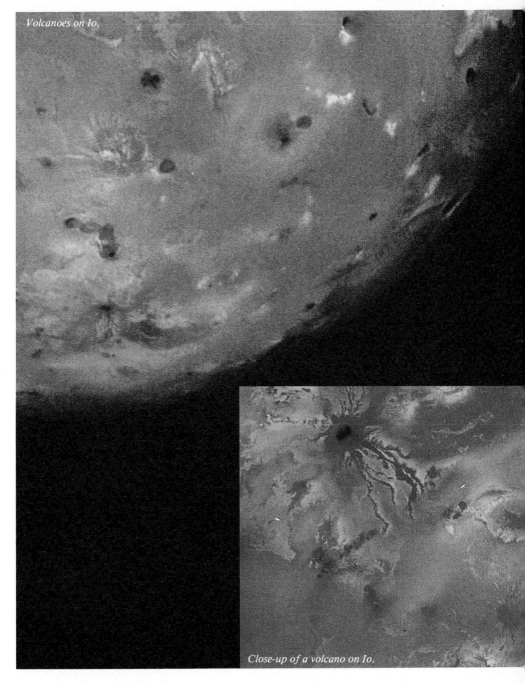

Volcanoes on Io.

Close-up of a volcano on Io.

Above *Venus — a Pioneer Orbiter view.*

Below *Map of Venus produced from Pioneer infra-red photographs showing low ground in blue-black rising to the yellow and red of mountains.*

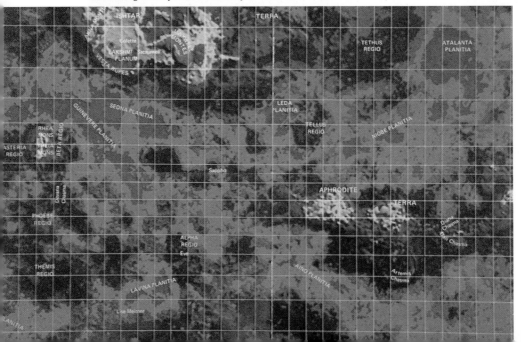

The first of the series was launched in April 1971.

Saros A period of 18 years 11.3 days, after which the Earth, Moon and Sun return to almost the same relative positions. It is due to the *regression of the nodes of the Moon's orbit. The Saros can be used to predict eclipses; it is usual for an eclipse to be followed by a similar eclipse 18 years 11.3 days later, though the slight differences from one Saros to another mean that the eclipses are not identical. For instance, the total solar eclipse of 1927 was seen from some parts of England, but the 'return' eclipse of 1945 was only partial from the British Isles.

Satellites Minor bodies which move round planets in the Solar System. So far as is known Mars has two satellites, Jupiter sixteen, Saturn twenty, Uranus five, Neptune two and the Earth and Pluto one each. Others have been suspected, and no doubt the outer giant planets have small satellites which await discovery. Data are given in the table.

There may be a basic difference between large satellites (eg Titan, Triton, the Galilean satellites of Jupiter and our Moon) and small bodies such as Phobos, Deimos and the outermost satellite of Saturn, Phœbe, which could possibly be captured asteroids.

The small outer satellites of Jupiter, Phœbe in Saturn's system, and Triton in Neptune's have *retrograde motion. Of these only Triton is large and presumably non-asteroidal. The fact that Triton is considerably more massive than *Pluto has been taken as an indication that Pluto was once a member of the Neptunian system, and that it was broken free by some external force, Triton being thrown into a retrograde orbit and *Nereid into an orbit with high eccentricity; but this is highly speculative. The satellites of Uranus have technically retrograde motion, but are not usually counted as such because they move in the same sense as the rotation of Uranus itself. It is generally assumed that non-asteroidal satellites were born at the same time as their primary planets, from material of the *solar nebula which was associated with these planets.

Satellite	Discoverer	Mean distance from primary in miles	Revolution Period in days	Orbital Eccentricity	Orbital Inclination	Diameter in miles	Magnitude
EARTH Moon	—	239,000	27.321	0.055	5.15	2,160	−12.7
MARS							
Phobos	Hall, 1877	5,800	0.319	0.021	1.1	12 × 14 × 17	11.6
Deimos	Hall, 1877	14,600	1.26	0.003	1.8	6 × 7 × 10	12.8
JUPITER							
Metis	Synnott, 1979	79,530	0.295	0.0	0.0	25	17.4
Adrastea	Synnott, 1979	80,160	0.298	0.0	0.0	15 × 12 × 10	18.9
Amalthea	Barnard, 1892	113,000	0.498	0.003	0.45	168 × 106 × 93	14.1
Thebe	Jewitt, Danielson, 1979	138,000	0.675	0.003	0.9	68 × 56	15.5

Satellite	Discoverer	Mean distance from primary in miles	Revolution Period in days	Orbital Eccentricity	Orbital Inclination	Diameter in miles	Magnitude
Io	Galileo and Marius, 1610	262,000	1.769	0.004	0.04	2,256	5.0
Europa		417,000	3.551	0.009	0.47	1,950	5.3
Ganymede		666,000	7.155	0.002	0.21	3,270	4.6
Callisto		1,170,000	16.689	0.007	0.51	2,983	5.6
Leda	Kowal, 1974	6,895,000	238.7	0.148	26.1	6	20.2
Himalia	Perrine, 1904	7,135,000	250.6	0.158	27.6	112	14.8
Lysithea	Nicholson, 1938	7,284,000	259.2	0.107	29	12	18.4
Elara	Perrine, 1904	7,295,000	259.7	0.207	24.8	50	16.7
Ananke	Nicholson, 1951	13,176,000	631	0.17	147	12	18.9
Carme	Nicholson, 1938	14,046,000	692	0.21	164	19	18.0
Pasiphaë	Melotte, 1908	14,605,000	735	0.38	145	25	17.7
Sunope	Nicholson, 1914	14,730,000	758	0.28	153	19	18.3
SATURN							
Atlas	From Voyager, 1980	85,580	0.602	0.002	0.3	24 × 16	18.0
Prometheus	From Voyager, 1980	86,640	0.613	0.004	0.0	87 × 62 × 46	16.5
Pandora	From Voyager, 1980	88,070	0.629	0.004	0.1	68 × 52 × 41	16.0
Janus	Dollfus, 1966	94,140	0.695	0.007	0.1	137 × 118 × 99	14.5
Epimetheus	Dollfus, 1966	94,110	0.694	0.009	0.3	87 × 71 × 62	15.5
Mimas	Herschel, 1789	115,300	0.941	0.02	1.52	244	12.9
Enceladus	Herschel, 1789	147,950	1.37	0.004	0.02	311	11.8
Tethys	Cassini, 1684	183,140	1.888	0.0	1.86	659	10.3
Telesto	From Voyager, 1980	183,140	1.888	0.0	1.86	15 × 14	19.0
Calypso	From Voyager, 1980	183,140	1.888	0.0	1.86	19 × 15 × 10	18.5
Dione	Cassini, 1684	234,570	2.737	0.002	0.02	696	10.3
Helene	From Voyager, 1980	234,570	2.737	0.005	0.2	22 × 19	18.5
Rhea	Cassini, 1672	327,560	4.518	0.001	0.35	951	9.7
Titan	Huygens, 1655	759,390	15.495	0.029	0.33	3201	8.4
Hyperion	Bond, 1848	920,510	21.277	0.104	0.43	218 × 145 × 124	14.2
Iapetus	Cassini, 1671	2,213,360	79.331	0.028	7.52	907	10.2-11.9
Phœbe	Pickering, 1898	8,050,960	550.4	0.163	175	137	16.5
URANUS							
Miranda	Kuiper, 1948	81,100	1.414	0.0	0.0	310	16.5
Ariel	Lassell, 1851	119,200	2.52	0.003	0.0	827	14.4
Umbriel	Herschel, 1802	166,070	4.144	0.004	0.0	750	15.3
Titania	Herschel, 1787	272,220	8.706	0.002	0.0	1,020	14.0

Satellite	Discoverer	Mean distance from primary in miles	Revolution Period in days	Orbital Eccentricity	Orbital Inclination	Diameter in miles	Magnitude
Oberon	Herschel, 1787	364,390	13.463	0.001	0.0	1,013	14.2
1986 U7		30,640	0.330			15	
1986 U8		33,126	0.372			20	
1986 U9		36,730	0.433			30	
1986 U3		38,377	0.463			30	
1986 U6	From Voyager 2,	38,968	0.475			45	
1986 U2	1985-6	39,993	0.493			45	
1986 U1		41,075	0.513			55	
1986 U4		43,455	0.558			30	
1986 U5		46,674	0.622			30	
1985 U1		53,382	0.762			90	
NEPTUNE							
Triton	Lassell, 1846	219,390	5.877	0.0	159.9	2,200?	13.5
Nereid	Kuiper, 1939	3,455,560	359.881	0.749	27.2	250?	19.0
PLUTO							
Charon	Christy, 1978	11,800	6.39	0.0	0.0	600?	±16.0

Satellites, Artificial See *Artificial Satellites.

Saturn The sixth planet in order of distance from the Sun. Apart from Jupiter, it is much larger and more massive than any other planet, but its overall density is less than that of water. It has been said that if Saturn could be dropped into a vast ocean, it would float!

Seen through a telescope, Saturn shows a yellowish globe which is obviously flattened, and is crossed by belts which are less marked than those of Jupiter. Occasional spots are seen. The famous white spot of 1933, discovered by the British amateur W. T. Hay (Will Hay, the actor) was the most prominent of modern times, but it did not persist for long.

The *Voyager space-craft which bypassed Saturn in 1980 and 1981 have increased our knowledge of the planet tremendously. It is now believed that there is a rocky core about three times as massive as the Earth, surrounded by a layer of liquid metallic hydrogen which is in turn overlaid by a layer of liquid molecular hydrogen and then the 'atmosphere', made up chiefly of hydrogen, with about 6 per cent of helium and smaller quantities of other elements. The temperature at Saturn's core may be as high as 15,000°C. Like Jupiter, Saturn sends out more energy than it would be expected to do if it depended entirely upon what it receives from the Sun, but the cause is different. Unlike Jupiter, Saturn has had ample time to lose the original heat produced during its formation, and the excess energy is more probably due to the effect of helium droplets sinking toward the centre of the globe through the lighter hydrogen. There is a magnetic field, 1,000 times stronger than that of the Earth but twenty times weaker than Jupiter's; the magnetic axis is within one degree of the axis of rotation.

The Voyagers showed many details on the surface of the planet, including some reddish and brownish spots. Windspeeds are high. The lower temperature as compared with Jupiter means that ammonia crystals form at higher levels, producing a 'haze' and giving the disk a rather bland appearance.

Comparative sizes of Saturn and the Earth.

Of course the most famous feature of Saturn is the ring-system, which is a glorious sight in even a small telescope when the rings are suitably tilted (as during the mid-1980s). The rings are made up of large numbers of icy particles, moving round Saturn in the manner of moonlets. It has been suggested that they are the débris of a former satellite, but it is more likely that they were simply material 'left over', so to speak, when Saturn itself was formed; moreover they lie inside the *Roche limit. Though the system measures 169,000 miles from one side to the other, the rings are very thin—perhaps less than a mile thick—so that when edge-on to the Earth, as in 1980, they almost disappear.

There are three main rings: A and B, which are separated by a gap known as the Cassini Division in honour of its discoverer (*G. D. Cassini) and C, an inner, transparent ring otherwise known as the *Crêpe or Dusky Ring. Ring A contains a minor division, discovered by J. F. *Encke and named after him. It had been thought that the rings were more or less homogeneous, and that the Cassini Division at least was an empty gap due to the gravitational perturbations of ring-particles by Saturn's satellites. However, the Voyagers showed that the rings are very complex. There are thousands of narrow ringlets and gaps, and there are narrow ringlets even within the Cassini Division. Moreover there are several rings

Saturn as seen from 34 million km by Voyager 1 in 1980 (JPL).

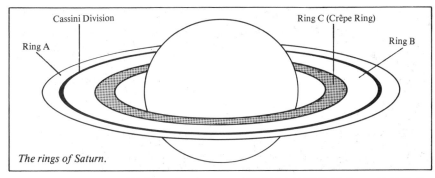

Ring A   Cassini Division   Ring C (Crêpe Ring)   Ring B

The rings of Saturn.

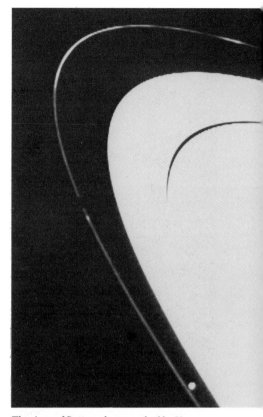

Dramatic 'close-up' of Saturn's rings photographed from 3.3 million km by Voyager 2 in 1981 (JPL).

The rings of Saturn photographed by Voyager 1, which also discovered the planet's four-teenth satellite, seen here just inside the 'F' ring (JPL).

outside the bright system, one of which (Ring E) extends out to the region between the orbits of the satellites Mimas and Enceladus. An inner ring region (Ring D), closer-in than the Crêpe Ring, is not observable from Earth and may not be a well-defined ring. Some of the rings depart from the strictly circular form, and one, Ring F, seems to be 'braided'; there are also strange, spokelike features in Ring B (much the brightest of the rings) which persist for some hours after emerging from the shadow of the globe. It cannot be said that as yet we fully understand the dynamics of the ring system, but it is significant that Ring F is stabilized by

two small satellites, Prometheus and Pandora.

Saturn has a wealth of satellites, but the system is quite unlike that of Jupiter. There is one very large satellite (*Titan); four of medium size, the *Rhea/*Iapetus pair and the *Dione/*Tethys pair; another pair of still smaller size, *Mimas/*Enceladus; the irregular *Hyperion; and the outer *Phœbe, which has *retrograde motion and a non-captured rotation, and may well be an ex--asteroid. The remaining satellites have been discovered fairly recently. In 1966 A. Dollfus observed an inner satellite which was named Janus; it later transpired that

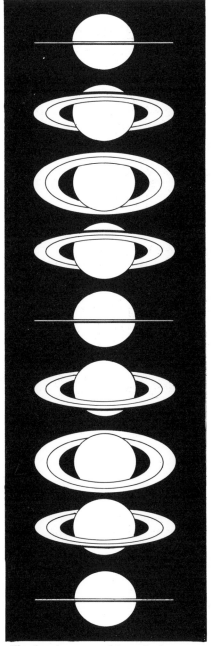

The changing aspect of Saturn's rings.

there are two co-orbital satellites, now known as Janus and Epimetheus and probably fragments of an originally larger body. Tethys has two co-orbitals, *Calypso and *Telesto; Dione has one, and a co-orbital satellite of Mimas has also been reported. No doubt other small satellites await discovery.

Saturn Nebula NGC 7009, a bright planetary nebula in Aquarius. Its name is derived from the dusty band crossing it, producing an appearance which slightly resembles that of Saturn. It is a very easy telescopic object, though photographs taken with large telescopes are needed to bring out the details.

Scheiner, Christoph (1573-1650) German astronomer, who was one of the first to make systematic observations of sun-spots.

Schiaparelli, Giovanni Virginio (1835-1910) Italian astronomer, best remembered for his observations of planets (in 1877 he was the first to describe in detail the now-discredited *canal system) and for his outstanding work upon cometary and meteoric studies.

Schmidt, Bernhard (1879-1935) Estonian inventor of the *Schmidt telescope.

Schmidt camera See *Schmidt telescope.

Schmidt, Julius (1825-1884) German astronomer, who spent much of his career as Director of the Athens Observatory in Greece. He was concerned mainly with the Moon, and drew up a lunar map which was much the best of its time.

Schmidt telescope (or Schmidt camera) A type of telescope invented in 1930 by Bernhard *Schmidt. It uses a special glass correcting plate, near the top of the tube, as well as a mirror; the mirror itself is spherical rather than parabolic. With a Schmidt, relatively wide areas of the sky may be photographed with one exposure; the definition remains good right up to the edge of the field. Instruments of this kind are purely photographic, but have proved to be of immense value. The 48-in

Schmidt at Palomar was used for the *Palomar Sky Survey. (The given aperture refers to the diameter of the correcting plate.)

Schönfeld, Eduard (1818-1891) German astronomer, who acted as assistant to F. W. *Argelander in the compilation of a very important star catalogue, the *Bonner Dürchmusterung*, subsequently extending it on his own account.

Schrödinger Rima A major valley on the Moon, lying just on the averted hemisphere and therefore not visible from Earth.

Schröter, Johann Hieronymus (1745-1816) German amateur astronomer, by profession Chief Magistrate at Lilienthal, near Bremen, where he had his private observatory. He was an outstanding planetary observer, and the real founder of *selenography. His observatory was destroyed by the invading French army, with the loss of many of his unpublished observations.

Schröter effect When the planet Venus is at half-phase, it is said to be at *dichotomy. Oddly enough, theoretical predictions do not usually agree with observation; when an evening object, and therefore waning, dichotomy is too early, while at morning elongations, when Venus is waxing, dichotomy is late. The effect may amount to several days; it is due to the effects of Venus' dense atmosphere. It was first noted by J. H. *Schröter. I was originally responsible for referring to it as the 'Schröter effect', but the term seems to have become generally accepted.

Schwabe, Heinrich (1789-1875) German amateur observer, who discovered the eleven-year sunspot cycle.

Schwarzschild, Karl (1873-1916) German astronomer who made fundamental contributions to studies of stellar evolution and distribution.

Schwarzschild radius The radius that a body must have if its *escape velocity is to be equal to the velocity of light. For example, consider the Sun, whose radius is approximately 432,000 miles. If the radius were reduced, without altering the mass, light would find it more and more difficult to escape; if the solar radius could be reduced to just below two miles, light would be unable to escape at all. We would, in fact, have a *Black Hole. With a body more massive than the Sun, the critical radius would naturally be greater. A body the mass of the Earth would need to be compressed to a diameter of only 1 cm to reach its Schwarzschild radius! The concept was first discussed by K. *Schwarzschild shortly before his death.

Schwassmann-Wachmann I Comet A periodical comet discovered by A. Schwassmann and A. A. Wachmann at the Hamburg Observatory in 1928. Its orbit lies almost entirely between those of Jupiter and Saturn, and has an eccentricity of only 0.11; it can thus be kept under constant observation. Normally it is very faint, but at times it shows outbursts which bring it temporarily within the range of small telescopes.

Scintar A type of *supernova remnant; see also *SS 433.

Scintillation The official term for 'twinkling'. It is due entirely to the Earth's atmosphere, and has nothing to do with the stars themselves. A star will twinkle much more strongly when low down than when high up, and may also flash various colours. (This is particularly noticeable with *Sirius, the brightest star in the sky.) Planets twinkle less obviously than stars, because they appear as small disks rather than as point sources.

Scorpius X-1 The first known cosmic X-ray source, discovered in 1962 by instruments carried aboard a rocket launched from the White Sands proving ground in New Mexico. In 1967 it was identified with a faint variable star, V.818 Scorpii, and is now known to be a binary system.

Seasons The Earth's axis is tilted at an angle of 23½° to the perpendicular (or, to put it another way, the equator is tilted by 23½° to the plane of the orbit). When the

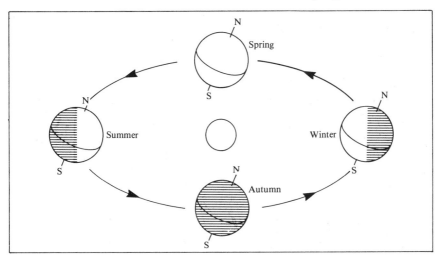

The seasons.

northern hemisphere of the Earth is inclined toward the Sun, it is summer in the northern part of the Earth and winter in the southern; six months later the positions are reversed. The Earth is at its closest to the Sun in northern winter (southern summer) but this makes very little difference, partly because the Earth's orbit does not depart widely from the circular form and partly because the greater area of sea in the southern hemisphere tends to stabilize the temperature. On Mars, where the axial tilt is much the same as ours and where northern winter also occurs near perihelion, the difference between the two hemispheres is much more marked; the Martian orbit is much more eccentric than that of the Earth, and there are no oceans. Therefore, the southern climate is much more extreme than the northern, with shorter, hotter summers and longer, colder winters.

Secchi, Angelo (1818-1878) Pioneer Italian spectroscopist, who divided the stars into well-defined spectral types.

Second of arc One-sixtieth of a *minute of arc.

Secular acceleration Because of friction produced by the *tides, the Earth's rotation is gradually slowing down; in other words, the day is becoming longer. The average increase daily amounts to only 0.00000002 second, but over a sufficiently long period of time the effects become noticeable. Another result of these tidal effects is that the Moon is receding from the Earth at a rate of about 4 cm per year.

As each day is 0.00000002 second longer than the previous day (allowing for slight random fluctuations), then a century (36,525 days) ago the length of the day was shorter by 0.00073 second. Taking an average between then and now, the length of the day was half this value, or 0.00036 second, shorter than at present. But since 36,525 days have passed by, the total error is $36.525 \times 0.00036 = 13$ seconds. Therefore the position of the Moon, when calculated back, will be in error; it will seem to have moved too far, ie too fast. This is the lunar *secular acceleration.* It shows up during calculations of *eclipses which took place in the distant past.

Secular variable A star which is suspected of a permanent change in brightness in recorded times. Thus *Denebola in Leo was ranked of the 1st magnitude by ancient astronomers, but is now just below the 2nd. All such cases are, however, very dubious because of the lack of precision of the old records.

Seeing The quality of the steadiness and clarity of the image of a celestial object. On a calm night, when there is little twinkling and the star does not seem to be moving about, the seeing is good; telescopically the disk of a planet will be sharp and steady. Seeing is often bad on a very transparent, brilliantly starlit night, when the images will dance about due to unsteadiness in the Earth's atmosphere.

Selenography The physical study of the Moon's surface.

Semi-regular variables Pulsating variable stars, generally red supergiants. Their magnitude ranges are much less than with the *Mira variables, and their periods are ill-defined. The brightest member of the class is *Betelgeux in Orion.

Serenitatis, Mare (The Sea of Serenity) One of the major lunar seas. It is crossed by ridges, and is comparatively regular in outline, with borders which are mountainous in places. The most prominent crater on it is Bessel; there is also the curious formation *Linné.

Sextant An instrument used for measuring the altitude of a celestial body above the horizon.

Seyfert galaxies Galaxies with relatively small, bright nuclei and weak spiral arms; attention was drawn to them by C. Seyfert in 1942. The spectra of the nuclei show emission lines, and there is evidence of high turbulence velocities, so that the nuclei are highly active; it is suggested that a Seyfert galaxy may contain a massive central black hole. Many Seyferts are strong radio sources; NGC 4151 is typical of this, and Seyferts also emit X-rays. About 10 per cent of all known highly luminous galaxies are Seyferts.

One famous Seyfert galaxy is M.77 in Cetus, which has a mass estimated at 800,000 million times that of the Sun. It is the most massive galaxy in *Messier's catalogue. The distance has been given as 52,000,000 light-years; the visual magnitude is about 9.

Shadow bands Wavy lines seen across the Earth just before and just after totality during a solar *eclipse. They are due to effects of the Earth's atmosphere. Shadow bands are not seen at every total eclipse, and are difficult to photograph well.

Shapley, Harlow (1885-1972) Great American astronomer, who was the first to give an accurate estimate of the size and shape of the Galaxy. He also made very important contributions to cosmology. In addition to his technical work he was the author of several excellent popular books.

Shaula The star Lambda Scorpii. For data, see *Stars. It is only just below the first magnitude, and is the brightest star in the Scorpion's 'sting'; it barely rises over any part of Britain. It appears to form a very wide pair with Lesath (Upsilon Scorpii), but there is no real connection between the two, since Lesath is much the more luminous, and looks fainter only because it is so much further away.

Shell stars Certain hot white stars which are unstable, and are known to be surrounded by shells of tenuous gas quite invisible optically; the shells are detectable spectroscopically. Pleione, in the *Pleiades, is a good example of a shell star.

Shooting-star The popular name for a *meteor.

Shuttle Manned American recoverable space-craft. It has achieved much, though in January 1986 the 25th flight ended tragically when the Shuttle *Challenger* exploded, killing all seven crew members.

Sidereal clock A clock regulated so as to keep *sidereal time.

Sidereal day This is described under the heading *Day.

Sidereal month This is described under the heading *Month.

Sidereal period (or periodic time) The time taken for a body to complete one journey round the Sun, or a satellite to complete one orbit round a planet.

Sidereal time The local time reckoned according to the apparent rotation of the *celestial sphere. The sidereal time is 0 hours when the *First Point of Aries crosses the observer's *meridian—that is to say, when the First Point of Aries culminates. Since the sidereal day is slightly shorter than the solar day, the local sidereal time will not usually be the same as ordinary civil time. The 24-hour system is always used.

The sidereal time for any observer is equal to the *right ascension of an object which lies on the meridian at that time. Thus when *Rigel (RA 5 hours 14 minutes) lies on the meridian as seen from Bristol, the sidereal time at Bristol is 5 hours 14 minutes.

Sidereal year This is described under the heading *Year.

Siderite An iron *meteorite.

Siderolite A stony-iron *meteorite.

Siding Spring Observatory Major Australian observatory, near Coonabarabran in New South Wales. The main instrument is the 153-in Anglo-Australian Telescope; there is also the 48-in UK Schmidt telescope.

Silvering The deposition of a thin layer of silver on to a mirror, to make it more reflective.

Simeiz Vallis A major valley on *Mercury.

Singularity In theory, a point where matter becomes infinitely dense—or, more accurately, space and time are infinitely distorted. It is suggested that a singularity lies at the centre of a *Black Hole.

Sinope The ninth satellite of Jupiter. For data, see *Satellites.

Sirenum, Mare A dark marking on Mars; latitude 50°S, longitude 150°.

Sirius The brightest star in the sky: Alpha Canis Majoris. For data, see *Stars. It has a faint White Dwarf companion, only about 1/10,000 as luminous as Sirius itself, and with a period of fifty years. Ancient astronomers described Sirius as a red star, whereas it is now pure white, but it is most unlikely that there has been any real change. Sirius is often nicknamed the Dog-Star.

Sisyphus An Apollo asteroid: No 1866. It was discovered in 1972. The diameter is no more than 2 miles.

Skylab The first true American space-station, launched from Cape Canaveral in May 1973. Three crews manned it in succession, carrying out important scientific work of many kinds. The last Skylab astronauts—Carr, Gibson and Pogue—landed back on Earth on 8 February 1974, afer having spent a record 84 days on the station. Skylab finally decayed in the Earth's atmosphere in 1979, somewhat earlier than expected, and scattered fragments over a wide area of Western Australia, fortunately without causing any casualties.

Slipher, Earl C. (1883-1964) American astronomer at the *Lowell Observatory, noted for his excellent planetary photographs.

Slipher, Vesto Melvin (1875-1966) Brother of *E. C. Slipher. He was a leading stellar spectroscopist, and succeeded *Lowell as Director of the Lowell Observatory.

Smyth, William Henry (1788-1865) British naval officer, who rose to the rank of Admiral. He established a private observatory at Bedford, and made many general observations; he is best remembered for his book *A Cycle of Celestial Objects*.

Smythii, Mare Smyth's Sea, a detached lunar mare on the Moon's north-east limb, named in honour of Admiral *Smyth.

Skylab during the third manned mission (NASA).

Solar constant The unit for measuring the amount of solar radiation received by the Earth. It amounts to 1.94 calories per minute per square centimetre. (A calorie is the amount of heat needed to raise the temperature of 1 gramme of water by 1° Centigrade.)

Solar parallax The trigonometrical *parallax of the Sun. It has a value of 8.79 seconds of arc, giving a mean distance for the Sun of 92,957,209 miles.

Solar System The system made up of the Sun, planets, satellites, comets, minor planets, meteoroids and interplanetary dust and gas. It is a very small part of the universe, and seems important to us only because we happen to live inside it.

The age of the Earth is about 4,700 million years; the Sun is presumably older, so that the Solar System itself must date back for at least 5,000 million years.

For a long time Laplace's *Nebular Hypothesis of the origin of the planets was accepted, but when fatal mathematical objections led to its rejection an alternative idea was put forward by Chamberlin and Moulton, of the United States. This involved the close approach of a wandering star, which tore material out of the Sun; when the passing star moved away, the torn-off material condensed into planets. The theory was developed by Sir James *Jeans, and became popular for a while, but it had so many weak links that it had to be given up. Various other ideas were proposed — for instance by Sir Fred *Hoyle, in which the planets were produced from a former companion of the Sun which exploded as a supernova — but it now seems that the planets built up by *accretion from a *solar nebula, a cloud of dust and gas associated with the youthful Sun. No doubt other stars have planetary systems of the same type.

Solar time, Apparent The local time, reckoned according to the Sun. Noon occurs when the Sun crosses the observer's *meridian, and is therefore at its highest in the sky.

Solar tower A tower with a *cœlostat at the top. The cœlostat reflects the Sun's light vertically downward, where it may be studied with spectrographic equipment. The solar tower is a convenient arrangement, since the main equipment can remain in a fixed position. However, the world's largest solar telescope, at *Kitt Peak, has an inclined tunnel rather than a vertical tower.

Solar wind A constant flow of electrified particles, streaming out from the Sun in all directions. It was discovered by instruments on the *Mariner 2 space-craft sent to Venus in 1962. The solar wind is not steady; it is 'gusty', particularly when the Sun is near the peak of its cycle of activity, and it is detectable out to great distances, as has been shown by results from the *Pioneer 10 and 11 probes now on their way out of the Solar System. The solar wind makes itself evident in many ways; for example it affects gas tails of comets, driving them outward from the nucleus in a direction away from the Sun.

Solstices The times when the Sun is at its northernmost point in the sky, declination 23½°N, around 22 June (summer solstice, midsummer in the northern hemisphere) and at its southernmost point, declination 23½°S, around 22 December (winter solstice: midwinter in the northern hemisphere). The actual dates of the solstices vary somewhat, because of the calendar irregularity due to *Leap Years.

Solstitial colure The *hour circle passing through the *solstices. It is described under the heading *Colures.

Solstitial points The points along the *ecliptic when the Sun reaches its maximum declination north or south (23½°) in either hemisphere. Obviously, the solstitial points are 90° away from the equinoxes.

Sombrero Hat Galaxy A nickname for the spiral galaxy M.104, in Virgo. It is of the 8th magnitude; the distance is about 41,000,000 light years. M.104 is distinguished by the dark streak crossing the nucleus — hence its popular name.

Somnii, Palus The Marsh of Sleep. A lunar sea between the Mare Tranquillitatis and the bright crater Proclus. Proclus is a ray-crater; two of its rays mark the limits of the Palus Somnii.

Somniorum, Lacus The Lake of the Sleepers. A mare area on the Moon, joining the Mare *Serenitatis to the Mare *Frigoris. The imposing crater Posidonius lies on its border.

Southern Cross *Crux Australis, the most famous of all the southern constellations; three of its stars are above the second magnitude. Crux, more kite-shaped than cruciform, is the smallest constellation in the sky, but also one of the richest. It contains the dark *Coal Sack nebula and the lovely *Jewel Box cluster, Kappa Crucis. The declination is around -60°, so that it is visible for most of the time from countries such as South Africa and Australia; it never rises over Europe.

Soyuz Manned Russian space-craft; the first of the series was launched in 1967. Soyuz vehicles ferry crew members and supplies to and from the *Salyut space-stations.

Spacelab Orbiting laboratory, carried by and contained in the *Shuttle. It enables scientists to work in orbit under really tolerable conditions.

Specific gravity The density of a substance compared with that of an equal volume of water. That of the Earth is 5.5; that of Saturn is only 0.7, so that overall Saturn is less dense than water.

Speckle interferometry A relatively modern technique, in which many very short-exposure pictures are taken with a powerful telescope and then treated electronically, thereby 'freezing out' the effects of turbulence in the Earth's

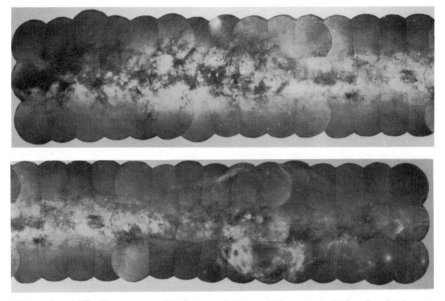

The Southern Milky Way, a composite photograph (Steward Observatory, University of Arizona).

atmosphere. It has already given excellent results.

Spectrogram A photograph of a *spectrum.

Spectroheliograph An instrument used for photographing the Sun in the light of one particular wavelength. If adapted for visual use, it is known as a *spectrohelioscope.*

Spectroscope An instrument used to split up the light of a luminous object. Equipment based upon the principle of the spectroscope has given us all our basic knowledge about the nature and composition of the stars and star-systems.

The actual splitting-up of the light into a *spectrum* is carried out by means of a *prism or a *diffraction grating. Basically, the idea is straightforward enough, but an astronomical spectroscope is extremely complex, and has to be used in conjunction with a telescope. Moreover, the telescope must be large; otherwise there will be too little light for useful analysis. (Matters are of course much easier with the Sun, where there is plenty of light available.)

According to *Kirchhoff's Laws, an incandescent solid, liquid or high-density gas will produce a *continuous* spectrum—a rainbow of colours from red at the long-wave end of the band through to violet at the short-wave end. A gas at lower density yields an *emission* spectrum of disconnected bright lines, each of which is characteristic of a particular element or group of elements. No element can duplicate the lines of another. This makes it possible to tell which substances are giving out the light, though in practice identifications are not always easy, since any element may produce a vast number of lines.

The Sun's bright surface or *photosphere gives a continuous spectrum. Above the photosphere is the *chromosphere, which is made up of gas at low density and would therefore yield a bright-line spectrum but for the presence of the rainbow background, which causes the bright lines to appear dark and produce an *absorption* spectrum. This effect was first interpreted by Josef von Fraunhofer, which is why the lines are often still called *Fraunhofer lines. Their positions and intensities are not affected, so that identifications can be made. One example will show what is meant. By itself, sodium vapour will produce two bright yellow lines (as well as many others). In the spectrum of the Sun there are two dark lines in the yellow part of the rainbow, agreeing in position with the sodium lines produced in the laboratory—so we can tell that there is sodium in the Sun.

Ordinary stars show spectra of the same basic type, though details differ; hot white stars have spectra which are easily distinguished from those of cooler red stars. Gaseous *nebulæ yield emission lines, the spectra of external *galaxies are something of a jumble, because they are made up of the combined spectra of many millions of stars, but the main absorption lines can still be identified.

Comparing the spectra of different stars makes it possible to work out the luminosities of the stars concerned, and this in turn gives a clue as to their distances from us. Measures of this kind are known as *spectroscopic parallaxes*, but the term is not a good one, because there is not to-and-fro shifting in position as with trigonometrical *parallax.

Another vitally important study is concerned with the *radial velocities of bodies, according to the *Doppler effect. All our ideas about the expansion of the universe depend upon this principle.

So far as the Moon and planets are concerned, the spectroscope is rather less effective, because these bodies shine by reflected sunlight, and produce spectra which are basically enfeebled spectra of the Sun. However, the light from a planet with an atmosphere has to pass through that planet's atmosphere twice before reaching us—once from Sun to planet, once from planet to Earth—and the substances in the atmosphere of the planet leave their imprint. For example, Venus was known to have a great deal of atmospheric carbon dioxide long before the first space-craft were sent there.

Spectroscopic binary A *binary system whose components are too close together to be seen separately. If they are in fairly quick motion round their common centre of gravity, one component will be approaching us while the other is receding; the first component will show a violet shift in its spectrum (Doppler effect) while the receding star will show a red shift. Consequently, the dark lines seen in the spectrum will appear double, becoming single again when the two components are moving *transversely and therefore are neither approaching nor receding. If only one component has a spectrum bright enough to be seen the lines will oscillate to and fro around a mean position. Thousands of spectroscopic binaries are now known.

Spectroscopic parallax This is described under the heading *Spectroscope.

Speculum The main mirror of a *reflector. Older mirrors were made of an alloy known as speculum metal; modern mirrors are of glass, either silvered or aluminized. Pyrex is a favourite material for telescope mirrors.

Principle of the spectroscope.

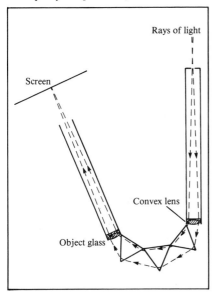

Rays of light

Screen

Convex lens

Object glass

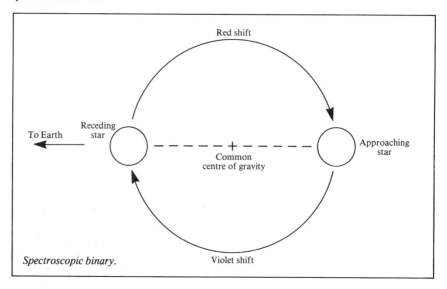

To Earth

Red shift

Receding star

Common centre of gravity

Approaching star

Violet shift

Spectroscopic binary.

Speculum metal An alloy of copper and tin, used for telescope mirrors before glass could be worked with sufficient accuracy. The last really large speculum-metal mirror was the 72-in made by the third Earl of *Rosse in 1845.

Spherical aberration The blurred appearance of an image as seen in a telescope, due to the fact that the lens or mirror does not bring the rays falling on its edge and on its centre to exactly the same focus. If the spherical aberration is really noticeable, then the optics of the telescope are defective, and should be reworked.

Spica The star Alpha Virginis; for data, see *Stars. It is an excessively close binary; the components are rather unequal, and are separated by no more than 11 million miles.

Spicules Jets, up to 10,000 miles in diameter, in the Sun's *chromosphere. They last for only four to five minutes each, and are associated with the solar granules. Spicules may be observed by means of a *filter.

Spörer's Law This is described under the heading *Sun.

Spring tide A tide produced when the Moon and Sun are pulling in the same sense, ie, when the Moon is new or full, at *syzygy. Spring tides have nothing to do with the season of spring. They are markedly stronger than the *neap tides.

Spumans, Mare (The Foaming Sea) A small lunar sea north of *Petavius and east of the boundary of the Mare *Fœcunditatis.

Sputnik The original Russian term for an *artificial satellite. It was the launching of Sputnik 1, on 4 October 1957, which opened the Space Age.

Stadius A famous lunar 'ghost crater', east of Copernicus. Its walls are so low that they can barely be traced. Stadius is 44 miles in diameter; it may once have been imposing formation, but has been overwhelmed by the lunar lava.

Star A self-luminous gaseous body. It may also be defined as 'a sun', since our Sun is a normal star.

The stars are grouped into *constellations, and are lettered and numbered accordingly. For the most conspicuous stars, Greek letters are used; the brightest star in a constellation should be lettered Alpha, the second Beta and so on, though the strict sequence is not always followed (in Sagittarius, for example, both Alpha and Beta are dim, while the two brightest stars are Epsilon and Sigma). For fainter stars, catalogue numbers are used. The Greek alphabet is as follows:

α Alpha	η Eta	ν Nu	τ Tau
β Beta	θ Theta	ξ Xi	υ Upsilon
γ Gamma	ι Iota	ο Omicron	φ Phi
δ Delta	κ Kappa	π Pi	χ Chi
ε Epsilon	λ Lambda	ϱ Rho	ψ Psi
ζ Zeta	μ Mu	σ Sigma	ω Omega

Sputnik 1, which heralded the beginning of the space age.

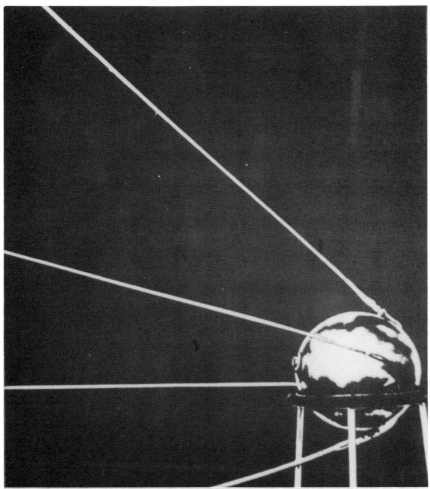

Many stars have individual or *proper* names, but these are now commonly used only for the brightest stars together with a few others of exceptional interest (such as *Mizar).

The *distances* of the stars are very great, and even *Proxima Centauri, our nearest stellar neighbour beyond the Sun, is more than four *light-years away. The distances of the closer stars are measured by trigonometrical *parallax; with more remote stars, less direct methods have to be used. The closest stars are as follows:

Star	Distance in light-years	Annual Proper per Motion in seconds of arc	Magnitude app	abs.	Luminosity Sun = 1	Spectrum
Proxima	4.2	3.75	10.7	15.1	0.0001	Me
Alpha Centauri	4.3	3.61	0.0	4.4	1.1	G4
			1.7	5.8	0.4	K5
Barnard's Star	5.8	10.27	9.5	13.2	0.0005	M5
Wolf 359	7.6	3.84	13.5	16.7	0.00002	M8
Lalande 21185	8.1	4.75	7.5	10.5	0.005	M2
Sirius	8.7	1.21	−1.4	1.4	26	A0
			8.5	11.4	0.003	dA

The only other naked-eye stars within twelve light-years of us are Epsilon Eridani (10.7), 61 Cygni (10.8), Epsilon Indi (4.7), *Procyon (11.4) and Tau Ceti (11.9).

Most of the naked-eye stars are much further away than this; in some cases (as with *Canopus) the distances and luminosities are very uncertain; those given here follow the Cambridge *Sky Catalogue*. Altogether there are fifty stars of magnitude 2.0 or brighter; in addition three variables (*Eta Carinæ, *Gamma Cassiopeiæ and *Mira Ceti) have exceeded the second magnitude at times. (This does not, of course, include novæ.) Data for the fifty brightest stars are given in the following table:

	Star	Name	Magnitude	Luminosity Sun = 1	Spec.	Distance in light-years
α	Canis Majoris	Sirius	−1.46	26	A1	8.7
α	Carinæ	Canopus	−0.72	200,000	F0	1,200
α	Centauri	—	−0.27	1.5	K1 + G2	4.3
α	Boötis	Arcturus	−0.04	115	K2	36
α	Lyræ	Vega	0.03	52	A0	26
α	Aurigæ	Capella	0.08	70	G8	42
β	Orionis	Rigel	0.12	60,000	B8	900
α	Canis Minoris	Procyon	0.38	11	F5	11.4
α	Eridani	Achernar	0.46	780	B5	85
α	Orionis	Betelgeux	var.	15,000	M2	310
β	Centauri	Agena	0.61	10,500	B1	460
α	Aquilæ	Altair	0.77	10	A7	16.6
α	Crucis	Acrux	0.83	3,200 + 2,000	B1 + B3	360
α	Tauri	Aldebaran	0.85	100	K5	68
α	Scorpii	Antares	0.96	7,500	M1	330
α	Virginis	Spica	0.98	2,100	B1	260

Star		Name	Magnitude	Luminosity Sun = 1	Spec.	Distance in light-years
β	Geminorum	Pollux	1.14	60	K0	36
α	Piscis Australis	Fomalhaut	1.16	13	A3	22
α	Cygni	Deneb	1.25	70,000	A2	1,800
β	Crucis	—	1.25	8,200	B0	425
α	Leonis	Regulus	1.35	130	B7	85
ε	Canis Majoris	Adhara	1.50	5,000	B2	490
α	Geminorum	Castor	1.58	45	A0	46
γ	Crucis	—	1.63	160	M3	88
λ	Scorpii	Shaula	1.63	1,300	B2	275
γ	Orionis	Bellatrix	1.64	2,200	B2	360
β	Tauri	Al Nath	1.65	470	B7	130
β	Carinæ	Miaplacidus	1.68	130	A0	85
ε	Orionis	Alnilam	1.70	26,000	B2	1,200
α	Gruis	Alnair	1.74	230	B5	69
ζ	Orionis	Alnitak	1.77	19,000	O9.5	1,100
ε	Ursæ Majoris	Alioth	1.77	60	A0	62
γ	Velorum	—	1.78	3,800	WC7	520
α	Ursæ Majoris	Dubhe	1.79	60	K0	75
α	Persei	Mirphak	1.80	6,000	F5	620
ε	Sagittarii	Kaus Australis	1.85	100	B9	85
δ	Canis Majoris	Wezea	1.86	130,000	F8	3,000
ε	Carinæ	Avior	1.86	600	K0	200
η	Ursæ Majoris	Alkaid	1.86	450	B3	108
θ	Scorpii	Sargas	1.87	14,000	G0	900
β	Aurigæ	Menkarlina	1.90	50	A2	72
α	Triangulum Australe	Atria	1.92	96	K2	55
γ	Geminorum	Alhena	1.93	82	A0	85
α	Pavonis	—	1.94	700	B3	230
δ	Velorum	Koo She	1.96	50	A0	69
α	Hydræ	Alphard	1.98	115	K3	85
β	Canis Majoris	Mirzam	1.98	7,200	B1	710
γ	Leonis	Algieba	1.99	60	K0 + G7	90
α	Ursæ Minoris	Polaris	1.99	6,000	F8	680
α	Arietis	Hamal	2.00	96	K2	85

No telescope will show a star as a measurable disk, and we have to depend for our basic information upon the *spectroscope. The stars show a tremendous range in luminosity and temperature, though considerably less in mass. The *Main Sequence stars, such as the Sun, are conventionally classed as dwarfs; the giant branch lies to the upper right of the *Hertzsprung-Russell Diagram, while the *White Dwarfs lie to the lower left. Spectral characteristics are as follows:

Type	Surface temperature in °C	Typical Star	Comments
W	Up to 80,000	γ Velorum	Bright lines in spectra. Wolf-Rayet stars: WN (nitrogen sequence) WC (carbon sequence). Unstable; highly luminous.

Type	Surface temperature in °C	Typical Star	Comments
O	40,000-35,000	ζ Puppis	Bright and dark lines.
B	25,000-12,000	Spica	Helium lines dominant.
A	10,000-8,000	Sirius	Hydrogen lines dominant.
F	7,500-6,000	Polaris	Calcium lines prominent.
G	giants 5,500-4,200	Capella	Metallic lines numerous.
	dwarfs 6,000-5,000	Sun	
K	giants 4,000-3,000	Arcturus	Weak hydrogen lines; metallic lines
	dwarfs 5,000-4,000	τ Ceti	strong.
M	giants 3,400	Betelgeux	Complicated spectra with many bands
	dwarfs 3,000	Proxima	due to molecules.
R	2,600	S Apodis	Carbon bands prominent.
N	2,500	R Leporis	Strong carbon bands.
S	2,600	χ Cygni	Prominent bands of titanium oxide and zirconium oxide.

Generally, W, O and B stars are white or bluish; A, white; F, slightly yellowish; G, yellow; K, orange; and M, R, N and S red. Many red stars are intrinsically variable.

Near a star's centre, the temperature is very high—at least 14 million degrees in the case of the Sun, which is comparatively mild by stellar standards. In these regions, where the star is producing its energy, the atoms are of course completely *ionized.

The source of stellar energy is found in nuclear reactions. In the 'power-house' of a Main Sequence star, hydrogen is being changed into helium; it takes four hydrogen nuclei to form one nucleus of helium, a process which is accompanied by release of energy and loss of mass. (The Sun is losing mass at the rate of 4 million tons per second.) High-mass stars evolve much more quickly than those of lower mass. The Sun will last for at least 5,000 million years in almost its present form, but a very luminous star, such as *Rigel, can hardly remain brilliant for more than a few millions of years.

A star's career depends upon its initial mass. In each case the star begins by condensing out of interstellar material, presumably in a 'pressure wave' in the Galaxy; *Bok globules may be protostars not yet hot enough to shine. Moreover stars tend to be born in groups, which survive for a long period as stellar associations. Gaseous nebulæ, such as

M.42 in Orion's Sword, are undoubtedly stellar birthplaces.

Nuclear reactions do not begin until the star has contracted, by gravitation, and its core has reached a temperature of around 10 million degrees. If the protostar has less than one-tenth the mass of the Sun, the temperature will never reach this value, and nuclear reactions will never be triggered off; the star simply shines feebly until it loses all its energy, turning into a cold, dead globe.

With a star of between 0.1 and 1.4 solar masses, things are different. As the star shrinks and heats up, it starts to shine—at first unsteadily; this is the so-called *T Tauri stage. Eventually the surrounding dusty 'cocoon' is blown away, and the hydrogen-into-helium process begins as the star joins the Main Sequence. It remains here for an immensely long period, but eventually the supply of available hydrogen is used up; energy production stops, so that the core shrinks and heats up to a temperature of perhaps 100 million degrees. This is enough to make helium react; the outer layers of the star expand and cool, but the total luminosity rises sharply as the star becomes a red giant. Finally the star's outer envelope is thrown off, producing a *planetary nebula; the core itself becomes very small and incredibly dense, because all its constituent atoms are broken up and packed together with little waste of space.

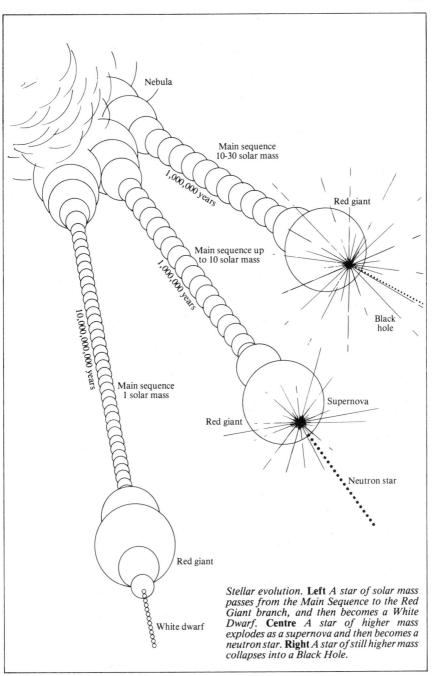

Nebula

Main sequence
10-30 solar mass

1,000,000 years

Red giant

Main sequence up
to 10 solar mass

1,000,000 years

Black
hole

10,000,000,000 years

Main sequence
1 solar mass

Red giant

Supernova

Neutron star

Red giant

White dwarf

Stellar evolution. **Left** *A star of solar mass
passes from the Main Sequence to the Red
Giant branch, and then becomes a White
Dwarf.* **Centre** *A star of higher mass
explodes as a supernova and then becomes a
neutron star.* **Right** *A star of still higher mass
collapses into a Black Hole.*

The star has reached the *White Dwarf stage, and after another very long period of feeble luminosity it presumably loses the last of its energy, becoming a dead 'black dwarf'.

A star of more than 1.4 times the mass of the Sun will evolve much more quickly, and will meet a more dramatic fate. Carbon nuclei are built up from helium nuclei, and at a temperature of some 700 million degrees heavier elements are built up such as neon, oxygen and silicon. There is a period when the star has an onion-like structure, with different reactions going on at different levels. When the temperature has soared to 3,000 million degrees or so, even iron nuclei are being created; but iron will not react further, and energy production stops. There is an 'implosion' (the opposite of an explosion) followed by a reaction, and the star disrupts itself in a *supernova outburst. The end product is a cloud of expanding gas, in the midst of which is the remnant of the old star; the protons and electrons are fused together to make neutrons, and the result is a *neutron star, far more dense even than a White Dwarf. Neutron stars rotate very rapidly, sending out pulsed radiation (*pulsars) but slow gradually down until they too have lost all their energy.

A star which is still more massive will not suffer a supernova outburst. When the final collapse starts, it is so sudden and so cataclysmic that nothing can stop it. Even the neutrons cannot resist the tremendous force, and the star creates a *Black Hole, inside which conditions are so beyond our experience that it is hard to visualize them. It has even been suggested that the old star may crush itself out of existence altogether!

This description of stellar evolution is very incomplete, and we cannot pretend to be certain of our facts with regard to the earliest and, particularly, the final stages of a star's career. One thing is certain: no star can last for ever. Material sent out by dying stars in supernova explosions contains relatively heavy elements, and these are included in the material used to form later generations of stars—such as our own Sun.

Star of Bethlehem The star mentioned in the Gospel according to St Matthew (Chapter 2) as having led the Wise Men to the place of the Nativity. Many theories to explain it have been proposed—the favourite being a conjunction of two or more planets appearing so close together that they merged into one exceptionally brilliant object. Unfortunately it has now been shown that no bright conjunctions were visible from the Holy Land at this time, and the idea is definitely untenable. It is not likely that the Star can have been a nova or supernova, as nothing of the sort was mentioned by contemporary astronomers, and in any case an outburst of this kind would remain visible for many nights. *Halley's Comet returned more than ten years too early, and it too would have been visible for some time. If the Star had been a familiar object, such as Venus, it would have caused no comment whatsoever. As we have nothing to guide us except the brief mention by St Matthew, we must admit that there is no scientific explanation for the Star of Bethlehem.

Star-streaming An effect of the rotation of the Galaxy, causing two main streams of stars which appear to move in opposite directions, one toward Orion and the other toward Ara. Attention was first drawn to the phenomenon in 1904 by J.C.*Kapteyn.

Starlink A system involving several identical computers, linked up. It enables observations to be 'played back' and analyzed in detail.

Steady-state theory The theory that the universe has always existed, and will exist for ever, so that it will always have much the same overall aspect. It was proposed in the late 1940s by a group of astronomers at Cambridge, and popularized by Sir Fred *Hoyle, but it has now been shown to be untenable, and has been rejected by almost all authorities. See also *Universe.

Stebbins, Joel (1878-1966) American astronomer who was the real founder of the study of *photoelectric photometry.

Steward Observatory Important observatory in Arizona, near *Kitt Peak. The main telescope is a 90-in reflector, completed in 1969.

Stickney The largest crater on Mars' satellite *Phobos. It is named after the wife of the discoverer of Phobos, Asaph *Hall; her maiden name was Stickney. It was she who persuaded Hall to continue the hunt for Martian satellites when he had been on the verge of giving up.

Stratosphere The layer in the Earth's atmosphere lying above the *troposphere. It extends from about 7 miles to about 40 miles above sea-level.

Strömgren sphere A symmetrical region of ionized gas, mainly hydrogen, surrounding a very hot star.

Struve, Friedrich Georg Wilhelm (1793-1864) German astronomer who spent much of his life in Russia, becoming Director first of the Dorpat Observatory and then of the new *Pulkova Observatory. He was concerned with stellar parallaxes, but is best remembered as a pioneer of double star astronomy.

Struve, Otto Wilhelm (1819-1905) Son of F. G. W. *Struve, who succeeded him as Director at Pulkova and was also a great double-star observer.

Struve, Otto (1897-1963) Great-grandson of F. G. W. *Struve. He was essentially a

Comparative sizes of the Sun and the Earth.

stellar spectroscopist; one-time Director of the *Yerkes Observatory and subsequently of the radio astronomy observatory at Green Bank, West Virginia.

Subdwarf A star which is between 1.5 and 2 magnitudes fainter than a normal Main Sequence star of the same spectral type.

Subgiant A giant star with a luminosity lower than that of a normal giant of the same spectral type.

Summer Triangle A completely unofficial name for the pattern made by *Vega, *Deneb and *Altair. I introduced the term in a *Sky at Night* television programme about 1959, and it now seems to be widely used. Of course it is inapplicable in the southern hemisphere, where the three stars are best seen in winter!

Sun The star which is the central body of the *Solar System. Its diameter is 865,000 miles; its volume is 1,300,000 and the mass 330,000 times that of the Earth. Its *specific gravity is 1.4.

The Sun is a *Main Sequence star of type G, and is producing its energy by the conversion of hydrogen into helium, losing mass at the rate of 4 million tons per second; but it is in no danger of imminent extinction, as it is at least 5,000 million years old, and is not likely to alter much for a very long period in the future.

It lies well away from the centre of the Galaxy, near the edge of a spiral arm; the distance between the Sun and the galactic centre is about 30,000 light-years. Of course the Sun shares in the general rotation of the Galaxy; its velocity is about 135

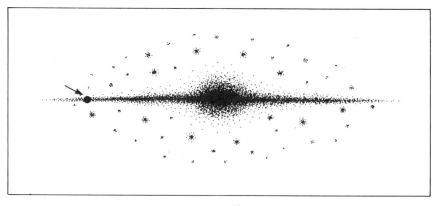

The Galaxy shown edge-on with globular clusters—position of the Sun arrowed.

miles per second so that it takes 225 million years to complete one revolution (a period which is sometimes called the 'cosmic year').

Telescopically, the Sun often shows dark patches known as *sunspots*. The spots look black, but are not really so: their temperature is about 2,000°C lower than that of the surrounding *photosphere, so that they appear dark by contrast. A large spot consists of a dark central part or *umbra*, surrounded by a lighter *penumbra*, though small spots may lack any penumbra. Spots tend to appear in groups, some of which are highly complex—though most major groups have two principal spots, a 'leader' and a 'follower'.

Sunspots are associated with strong magnetic fields. This was first established by G. E. *Hale in 1908. Hale also found

The structure of sunspots.

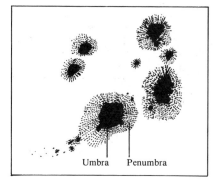

Umbra   Penumbra

that in a spot-group the leader and the follower are of opposite magnetic polarities—and that conditions are the same in this respect over a complete hemisphere of the Sun, though reversed in the opposite hemisphere. Small spots may persist for only a few hours, though large groups may remain visible for months before dying out. They are associated with the bright *faculæ, and active groups may produce the very energetic *flares which emit charged particles and radiation.

It cannot be claimed that our knowledge of the origin of sunspots is at all complete. According to a theory proposed by *H. Babcock in 1961, the Sun's magnetic lines of force run below the bright surface from one magnetic pole to the other. The Sun does not rotate as a solid body would do; the rotation is *differential—less than 25 days near the equator, as much as 34 days at the poles. This means that the magnetic lines are distorted and drawn out into loops. Over a period of years the lines are coiled right round the Sun and bunched near the poles. Finally a loop of magnetic energy erupts through the surface, cooling it and producing the two characteristic spots of a group. After about eleven years the knots have become so complex that they break; in returning to its original condition the Sun 'overshoots', so that the polarities of the spots in opposite hemispheres are reversed.

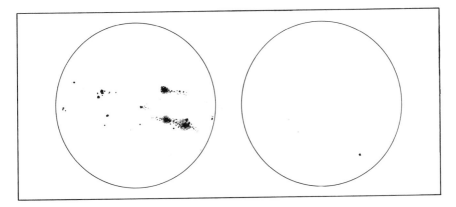

Active and quiet Sun.

The eleven-year cycle of activity was first discovered by H. *Schwabe. At spot-maximum there may be many groups on view, while during spot-minimum the disk may be spot-free for many consecutive days. The first spots of a new cycle appear in moderately high solar latitudes; as the cycle progresses, the spots are formed nearer and nearer to the equator (never quite reaching it), while the first spots of a new cycle may appear before the last spots of the old cycle have died away. This is known as *Spörer's Law, in honour of its discoverer. The cycle is not regular, and eleven years is only an average; moreover, there is good evidence that between 1745 and 1815 the cycle was suspended, so that virtually no spots were seen (the so-called *Maunder Minimum).

No part of the Sun's surface is calm.

The solar cycle. Note the Maunder Minimum (1645-1715) when the cycle was apparently suspended.

The photosphere has a granular structure; each granule is several hundreds of miles in diameter, and persists for less than ten minutes. It is estimated that the whole surface includes about four million granules at any one time, while *spicules rise from the photosphere.

Spectroscopically, over seventy elements have now been identified in the Sun. By far the most abundant are hydrogen (71 per cent) and helium (27 per cent), all the rest making up only 2 per cent. There is steady emission of *solar wind. The Sun is a source of *cosmic rays, *X-rays and *radio waves; it is also a source of *neutrinos, though the number of neutrinos appears to be much less than theory predicts (see also *Homestake Mine). Above the photosphere come the *chromosphere (containing the *prominences) and the *corona, which is of tremendous extent. With the naked eye, these regions can be seen only during a total solar *eclipse.

It is extremely dangerous to look directly at the Sun through any optical instrument, even when a dark filter is fitted. The only sensible way to observe

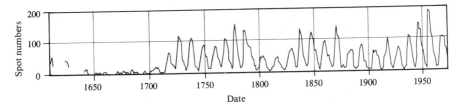

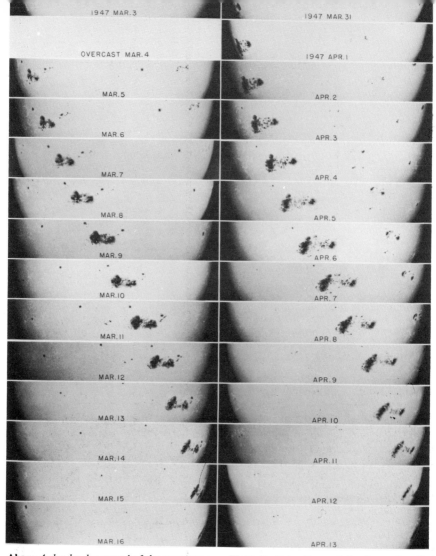

1947 MAR.3
OVERCAST MAR.4
MAR.5
MAR.6
MAR.7
MAR.8
MAR.9
MAR.10
MAR.11
MAR.12
MAR.13
MAR.14
MAR.15
MAR.16

1947 MAR.31
1947 APR.1
APR.2
APR.3
APR.4
APR.5
APR.6
APR.7
APR.8
APR.9
APR.10
APR.11
APR.12
APR.13

Above *A day-by-day record of the great sunspot group of 1947* (Mount Wilson and Palomar).

Right *Partial eclipse of the Sun, also displaying sunspots.*

sunspots is to use the telescope as a projector, sending the image on to a white screen held or fixed behind the eyepiece.

The Sun will not last for ever. Eventually it will swell out to become a red giant (see *Star), and this will involve the destruction of the Earth, since the luminosity will be about 100 times the present value and the size of the Sun's globe will be greater than the diameter of the orbit of the Earth. This will be succeeded by the White Dwarf stage, and final extinction.

Solar physics is a vitally important branch of modern astronomy, and some observatories concentrate entirely upon it. The Sun is the only star close enough to be examined in detail, and by studying it we are also learning more about the other stars, which we can see only as points of light.

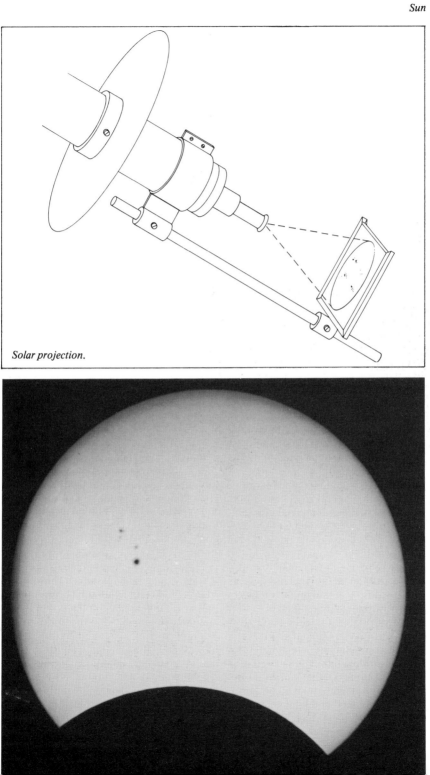

Solar projection.

Sundial An instrument used to tell the time, by using an inclined rod or plate (the *gnomon* or *style*) to cast a shadow on a graduated scale; the gnomon points to the celestial pole. The sundial shows apparent time; to obtain mean time the reading on the scale must be corrected to allow for the *equation of time.

Supergalaxy It is known that galaxies occur in clusters. A cluster of clusters may make up what is termed a supergalaxy or supercluster. Our *Local Group and the *Virgo Cluster may make up part of such a system.

Supergiant stars The most luminous of all

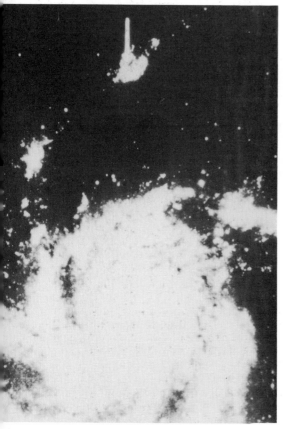

Supernova in Messier 101.

types of stars, usually with *absolute magnitudes of from −5 to −9.

Superior planets The planets beyond the orbit of the Earth in the Solar System— that is to say, all the principal planets apart from Mercury and Venus.

Supernova A tremendous stellar explosion, involving the virtual destruction of a star. Supernovæ are of two types:

Type I. Here we have a *binary system, of which one component is a Main Sequence star and the other is a White Dwarf. The White Dwarf pulls material away from its companion, and this material accumulates; when the mass of the carbon/oxygen core has become great enough, the carbon reacts, 'burning' in a wave which travels outward and completely disrupts the White Dwarf. The peak *absolute magnitude may be as high as −19. Following the outburst, it takes at least a year for the decline in brightness to be complete.

Type II. Here we have a metal-rich star much more massive than the Sun. When the final collapse starts, with the exhaustion of all nuclear processes, there is an implosion, followed by a shock-wave which reaches the star's surface and blows it away. The maximum absolute magnitude is about −17. The end product is a *pulsar or *neutron star, while the gas-cloud eventually dissipates.

Four supernovæ have been seen in our Galaxy during the past thousand years; those of 1006 (in Lupus), 1054 (in Taurus), 1572 (in Cassiopeia) and 1604 (in Ophiuchus). All these have become bright enough to be seen with the naked eye in broad daylight. The 1006 star was probably the most brilliant, and may have equalled the quarter-moon, but records of it are sparse; the 1054 star produced the *Crab Nebula; that of 1572, *Tyche's Star, is marked by a radio source but no pulsar, and the same is true of the 1604 star, *Kepler's supernova. It is thought that another supernova occurred about 1702 in Cassiopeia at a distance of around 11,000 light-years, but was not observed because it was hidden by interstellar material; a powerful radio source marks the site. Many galactic radio sources, such as that in *Vela, are also thought to have been due to supernovæ which blazed forth long ago.

On average, each galaxy of our kind seems to produce one observable supernova every 600 years or so, so that on this reckoning we are not yet due for another. Of course, there is no means of telling, and a supernova might be seen at any time; one possible future candidate is *Eta Carinæ. Astronomers would welcome the opportunity to study one with modern equipment. Meanwhile, many supernovæ have been seen in other galaxies, and one—*S Andromedæ, observed in 1885 in the Andromeda Galaxy—reached the sixth magnitude.

Surt An active volcano on *Io, observed from the *Voyager probes.

Surveyor probes Soft-landing American unmanned lunar vehicles. There were seven in all:

Surveyor 1 (1966). Successful landing in the *Oceanus Procellarum. In the following six weeks it returned over 11,000 pictures.

Surveyor 2 (1966). Failure; crashed near Copernicus.

Surveyor 3. (1967). Successful landing in the Oceanus Procellarum; returned over 6,000 pictures.

Surveyor 4 (1967). Failure; crashed in the Sinus *Medii.

Surveyor 5 (1967). Successful landing in the Mare *Tranquillitatis; returned over 18,000 pictures.

Surveyor 6 (1967). Successful landing in the Sinus Medii; returned over 30,000 pictures.

Surveyor 7 (1968). Successful landing on the outer wall of *Tycho. Over 21,000 pictures were returned, and chemical analyses of the surface material carried out.

Parts of Surveyor 3 were brought back by the astronauts of Apollo 12, who landed within walking distance of it.

Sutherland, Observatory at Now the main astronomical centre of South Africa; most of the main telescopes have been sent there, including the 74-in reflector from the now-dismantled Radcliffe

Sword-Handle cluster in Perseus (see next page).

Observatory near Pretoria. Sutherland lies in Cape Province, north of Cape Town, and is thought to have the best observing conditions in the Republic.

Swift One of the two main craters on *Deimos. It is named in honour of Jonathan Swift, who predicted that Mars would prove to have two satellites—though this prediction was hardly scientific; it was reasoned that since the Earth had one known satellite, and Jupiter four, Mars could hardly be expected to manage with less than two!

Swift, Lewis (1820-1913) American astronomer, who discovered thirteen comets and 900 nebulæ.

Swift-Tuttle Comet A periodical comet discovered in 1862, independently by L. *Swift and H. P. Tuttle. It reached the 4th magnitude, and developed a 10° tail. The period was calculated as being 120 years, and it was established that this is the parent comet of the *Perseid meteor stream. It was expected back about 1982, but so far it has not been seen. Unless it has disintegrated (which seems unlikely), either it has come and gone unseen, or else the calculated period was wrong, in which case the comet may be recovered in the fairly near future.

Sword-Handle Nickname of the double cluster in Perseus, H.VI.33-4. (Chi-h Persei). The clusters are in the same low-power telescopic field; they are just visible with the naked eye.

Symbiotic variables See *Z Andromedæ variables.

Synchronous rotation Alternative name for *Captured rotation.

Synchrotron radiation Radiation emitted by electrons moving at very high velocities in the presence of a magnetic field.

Synodic period The interval between successive *oppositions of a *superior planet. The periods for each planet are given in the table under the heading *Planets. Mars has much the longest synodic period, because on the astronomical scale it is not far beyond the Earth's orbit, and moves at a comparable speed. For the *inferior planets (Mercury and Venus) the term is applied to the interval between successive *inferior conjunctions with the Sun.

Syrtis Major The principal dark marking on *Mars, formerly known as the Kaiser Sea or the Hourglass Sea. It is V-shaped, and visible with a small telescope under good conditions; it was first recorded by C. *Huygens in 1659. It is now known to be a plateau rather than a depression.

Syzygy The position of the Moon in its orbit when new or full.

T

T Tauri stars Very young stars, which have started to shine but have not yet reached the *Main Sequence. They vary irregularly in light, and many are associated with nebulosity; for example the *Orion Nebula contains many of them. Their spectra are usually of types F, G or K. During the T Tauri stage, while contracting toward the Main Sequence, the star may eject considerable amounts of material in the form of a stellar wind. Sub-classes of these variables are named after their prototype stars YY Orionis, T Orionis and *RW Aurigæ.

Tachyons Particles which move faster than light. They are theoretical concepts, but whether they really exist or not is a matter for debate. If they do, it follows that they can never move as *slowly* as light!

Tarantula Nebula The great gaseous nebula 30 Doradûs, in the large *Magellanic Cloud. It is the largest known object of its type, and is a magnificent telescopic sight.

Tau Ceti One of the two nearest stars to bear any real resemblance to the Sun, though it is smaller and cooler. It may be a candidate as a centre of a planetary system, though proof is lacking. It was one of the targets for the unsuccessful *Ozma project. As its apparent magnitude is 3.5, it is easily visible with the naked eye.

Taurid Meteors A minor shower, reaching its maximum on 12 November, though the whole shower lasts from 20 October to 25 November. The *ZHR is usually about 12; the parent comet is *Encke's.

Tautenberg Observatory A major observatory in East Germany, near Jeha, known officially as the Karl Schwarzschild Observatory. The main telescope is a 72 in reflector, completed in 1960, which can be used as several different optical systems, and can quickly be converted from one system to another.

Taygete One of the main stars of the *Pleiades cluster.

Tebbutt, John (1834-1916) Australian amateur astronomer who discovered several comets, including the bright comets of 1861 and 1881.

Tektites Small glassy objects found in a few restricted areas on Earth, mainly in Australia, the Ivory Coast, Czechoslovakia and Texas and Georgia in the United States. No tektite has ever been found in Britain. They appear to be aerodynamically shaped, and to have been heated twice; their ages range from the geological Miocene to Late Pleistocene. Their nature is unknown. They may come from space; but it is equally likely that they have been sent out by terrestrial volcanoes. Suggestions that they may have come from the Moon do not seem to be very plausible.

Telescope The main instrument used to collect light from celestial bodies. There are two main types, the *refractor and the *reflector. The first refractor was made in Holland around 1608, and the first reflector, made by Isaac *Newton, was submitted to the Royal Society in 1671. Today, the main emphasis is upon reflectors rather than refractors. A new development will be the forthcoming launching of a major telescope into space; this is the 94-in *Hubble reflector. In the foreseeable future there will no doubt be telescopes operating from the surface of the Moon, which has no atmosphere, and where *seeing conditions will always be perfect.

Some telescopes owned by amateurs are highly sophisticated. However, it is probably true to say that the minimum useful size for an astronomical telescope is an aperture of 6 in for a reflector, or 3 in for a refractor. Binoculars are preferable to telescopes smaller than this.

A selection of tektites.

Telluric lines Spectral lines, seen in the spectra of celestial objects, which are due to gases in the Earth's atmosphere through which the light has to pass. They are therefore not *Doppler-shifted, which makes them readily identifiable.

Telstar The first successful television communications satellite. The initial transmission, between England and America as well as other countries, took place on 10 July 1962.

Tempel, Ernst Wilhelm Liebrecht (1821-1889) German astronomer who worked successively in Italy and France. He discovered the nebula in the *Pleiades, 5 asteroids and several comets.

Tempel I Comet A periodical comet discovered by the German astronomer E. W. *Tempel. It is always faint; the period is 5.5 years. It was studied from the *IRAS satellite in 1983, but no dust-tail was found.

Tempel II Comet Also discovered by Tempel; it is brighter than his first comet, and at times, as in 1925, can rise to above the 7th magnitude. The period is 5.3 years. In 1983, the *IRAS satellite discovered that it has a long dust tail, not visible in ordinary light.

Temporary stars An old term for *novæ; it is now obsolete.

Limb and terminator.

Terminator The boundary between the day and night hemispheres of the Moon or a planet. Since the lunar surface is mountainous, the terminator is rough and jagged, and isolated peaks may even appear to be detached from the main body of the Moon. The terminators of Mercury, Venus and Mars appear smooth as seen from Earth; though both Mercury and Mars have mountainous surfaces, the terminator irregularities are not obvious from long range. When the terminator bisects the disk, so that the Moon or inferior planet is an exact half-phase, it is said to be at *dichotomy.

Tethys The third satellite of Saturn. For data, see *Satellites.

Tethys is only slightly smaller than *Dione, but is much less massive, and seems to be made up largely of ice with an admixture of rock. There is one huge crater, Odysseus, with a diameter of 250 miles—larger than the diameter of Saturn's satellite *Mimas. It is not very deep, but it is clearly preserved, and if it had been caused by a meteoritic impact it seems that Tethys itself would have been in danger of disruption. There are many other craters, and a tremendous trench, Ithaca Chasma, over 1,200 miles long, running from the north pole across the equator and along to the region of the south pole. Its average width is 62 miles, and it is over 3 miles deep at maximum. Nothing quite like it is known elsewhere in the Solar System. Two small satellites,

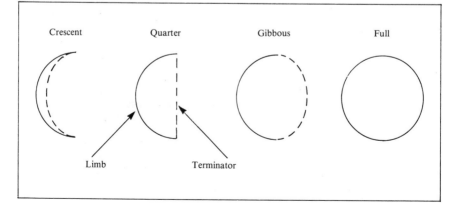

| Crescent | Quarter | Gibbous | Full |

Limb   Terminator

*Calypso and *Telesto, move in the same orbit as Tethys.

Thalassoid A lunar basin with a light floor. Thalassoids are found mainly on the region of the Moon which is permanently turned away from the Earth.

Thales (BC 640-560) Thales of Miletus was the first of the great Greek philosophers. His ideas seem very curious today, but he was an able observer, and is said to have made an accurate prediction of an *eclipse.

Tharsis Ridge The main volcanic region of Mars. It includes the highest Martian volcanoes: Olympus, Ascræus, Pavonis and Arsia.

Thatcher's Comet Discovered in 1861, when it just reached naked-eye visibility; it is the parent comet of the *Lyrid meteors. It is not due back yet awhile, as its period has been estimated as 415 years.

Thebe The fifteenth satellite of Jupiter, moving between the orbits of *Amalthea and *Io. For data, see *Satellites.

Theia Mons Active volcano in the *Beta Regio highland of Venus.

Themis A reported satellite of Saturn, announced by W. H. *Pickering in 1904, and said to move between the orbits of *Titan and *Hyperion. It has never been confirmed, and apparently does not exist. Asteroid No 24, discovered by de Gasparis in 1852, is also named Themis; it is one of the larger members of the swarm, with a diameter of about 145 miles. Its period is 5.6 years, and the mean opposition magnitude is 11.8.

Theophilus Great lunar crater, 64 miles across and 18,000 ft deep, with high terraced walls and a central mountain group. It lies on the border of the Mare *Nectaris, and is one of a chain of three huge walled formations; the others are Cyrillus and Catharina.

Thermocouple An instrument used for measuring very small quantities of heat.

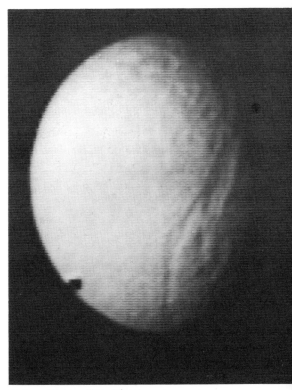

Tethys, third satellite of Saturn.

Basically, it consists of a circuit made up of wires of two different metals, joined together. If one of the joins is warmed, and the other kept at a constant temperature, an electric current is set up in the wire; the amount of the current is a key to the amount of heat involved. When used together with large telescopes, thermocouples are remarkably sensitive.

Thuban The star Alpha Draconis; magnitude 3.6. In ancient times, as when the Pyramids were built, Thuban was the north polar star. It is 230 light-years away, and is about 90 times as luminous as the Sun; it is a white star of type AO.

Tides The rise and fall of the ocean waters, due to the gravitational pulls of the Moon and (to a lesser extent) the Sun.

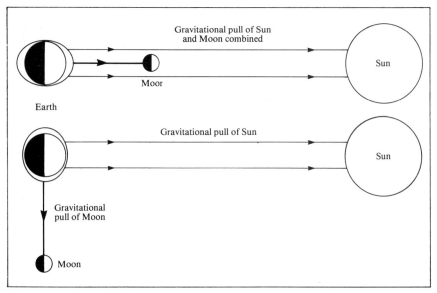

Spring (top) and neap tides.

If the Earth were surrounded by a uniform shell of water, the Moon would tend to heap up the water beneath it; as the Earth rotated on its axis the 'heap' would stay under the Moon, and would sweep right round the Earth once in 24 hours. Since there would also be a 'heap' on the far side of the Earth, each area would have two high tides per day. Actually, matters are not so straightforward as this, because the oceans are irregular both in outline and in depth, but the basic principles are clear enough.

When the Sun and Moon are pulling in the same sense, with the Moon at *syzygy, the tides are at their strongest (*spring tides). When the Moon is at *quadrature, it is pulling against the Sun, producing weaker (*neap) tides.

Time See under the headings *Sidereal time, *Solar time, *Greenwich Mean Time, *Universal time and *Equation of time.

Time dilation effect A consequence of the theory of *relativity. To a moving observer, time passes more slowly than it should do, compared with an observer at rest. The effect becomes evident only at very high velocities, comparable with that of light. Thus if an observer made a round trip to a star 12½ light-years away, moving at a steady rate of 0.999 the velocity of light, his journey would last for 25 years by Earth time—but on his own scale the period would be only one year. This leads on to the 'twin paradox'. If one twin went on such a journey, leaving the other twin behind, the traveller would on return seem much the younger of the two.

There is a practical test of this theory. *Cosmic rays entering the upper atmosphere create particles which are known as mu-mesons; these particles are very short-lived, and ought not to persist for long enough to reach the Earth's surface—but they do, because they are moving so fast that their time-scale, relative to ours, is slowed down. In 1974, too, it was found that a clock carried round the world in a high-speed jet aircraft 'ran slow' against an identical clock kept in a laboratory, though of course the difference was very slight indeed.

Titan The largest satellite of Saturn. For data, see *Satellites. As it is above the 9th magnitude, it is an easy telescopic object.

Apart from *Ganymede, Titan is the largest satellite in the Solar System, and is the only planetary satellite known to have a dense atmosphere. This atmosphere was detected spectroscopically by G. P. *Kuiper in 1944, but until the flight of *Voyager 1, which by-passed Saturn in November 1980, its composition was not known.

The Voyager results showed that Titan is a remarkable world. The atmosphere is made up chiefly of nitrogen, with a considerable quantity of methane; the ground pressure is 1.6 times that of the Earth's air at sea-level, and the clouds hide the surface of Titan permanently, so that the Voyager pictures showed only the upper layers. There are various organic compounds, giving the clouds a strong orange colour, and it has been said that all the ingredients for life exist, though the very low temperature has probably prevented life from appearing.

Titan has a *specific gravity of 1.9; according to one estimate the globe consists of about 55 per cent rock and the rest of ice. We know nothing definite about the surface conditions. The temperature is near that of the *triple point of methane; there may be cliffs of solid methane, rivers of liquid methane, and a constant rain of organics from the orange clouds! Alternatively, it has been suggested that there may be a methane or ethane ocean of great depth. We can hardly hope to find out more until a new space mission is sent there, to map the surface by radar and also dispatch an entry probe.

Tiros A weather satellite.

Titania The most massive satellite of *Uranus. For data, see *Satellites. It has always been thought to be the largest of the Uranian satellites; recent measurements indicate that it may be slightly smaller than *Oberon, though the two are very similar. Presumably Titania has an icy surface. We should find out more when the *Voyager 2 space-craft by-passes Uranus in January 1986.

Titius-Bode Law Alternative name for *Bode's Law.

Tombaugh, Clyde (1907-) Great American astronomer, who discovered *Pluto in 1930. He is now Professor Emeritus at the University of Las Cruces. His book *Out of the Darkness: The Planet Pluto*, in which he paid me the honour of asking me to collaborate, was published in 1980, half a century after the discovery of Pluto.

Toro Asteroid No 1685. It is of the *Apollo class; its distance from the Sun ranges between 71,500,000 miles and 182 million miles, and the period is 1.6 years. When discovered, by Wirtanen at *Lick in 1948, it was widely reported as being a minor satellite of the Earth. Of course this is quite wrong; it is an ordinary Apollo asteroid. It can at times come within 14 million miles of the Earth (as in 1980) and is one of the asteroids successfully contacted by *radar.

Tower telescope Equipment used for studying the Sun. Its chief advantage is

The cloud-covered surface of Saturn's satellite Titan as seen by Voyager 1 in 1980 (JPL).

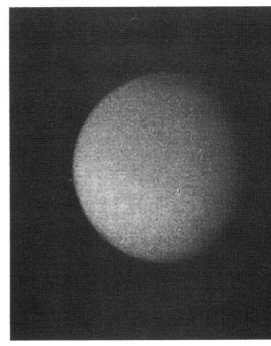

that the heavy analytical equipment need not be moved at all, since the sunlight is reflected in a fixed direction by the *cœlostat at the upper end of the tower.

Tranquillitatis, Mare (The Sea of Tranquility) A major sea joining the Mare *Serenitatis to the Mare *Fœcunditatis. It was here that the first lunar landing was made, from Apollo 11, in July 1969. There are no really large craters on the Mare Tranquillitatis.

Transfer orbit The most economical orbit for a space-craft sent to another planet. To move along the shortest path would need far too much propellant. Instead, the probe is put into an orbit which will swing it either in or out to the orbit of the target planet, so that the planet and the probe will meet. Almost all the journey is therefore carried out in *free fall, without the expenditure of fuel. Transfer orbits were once known as *Hohmann orbits*.

Transient lunar phenomena (TLP) Over many years, most serious lunar observers have reported elusive local obscurations and glows in certain areas of the Moon, notably near the brilliant crater *Aristarchus. These are now known as TLP (a term for which I believe I was originally responsible). Until 1958 the results were treated cautiously by professional astronomers, but then a TLP was observed in the crater Alphonsus by N. A. Kozyrev, at the Crimean Astrophysical Observatory, and a spectrum was obtained. Further conclusive TLP observations have been made since, and their reality is not longer seriously challenged, though interpretations differ; they are probably due to gases sent out from below the outer *regolith.

There is a close link between TLP and the mild ground tremors or *moonquakes recorded by the Apollo *Alseps; also, TLP are commonest near lunar *perigee, when the crust is under its greatest strain. The main TLP areas are the boundaries of the regular maria, or else regions crossed by *rills. The tremors are of course very slight by terrestrial standards, and present no threat to a future Lunar Base.

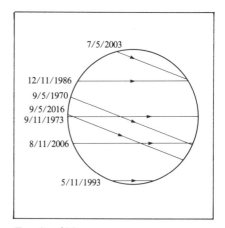

Transits of Mercury.

Transit The passage of a celestial body across the observer's *meridian. Thus the *First Point of Aries must transit the meridian at 0 hours sidereal time.

There are several other uses of the term. Mercury and Venus are said to transit when they pass between the Earth and the Sun, so that the planet shows up as a black disk against the solar surface; transits of Mercury are not particularly uncommon (see *Mercury) but the next transit of Venus will not take place before the year 2004. Transits of the four *Galilean satellites of Jupiter are easy to follow with a small telescope; the satellites and their shadows may be followed as they pass across the Jovian disk.

As a planet spins on its axis, various markings are brought to the central meridian as seen from Earth, and are then said to be in transit. Observations of the transits of features on Jupiter have provided us with most of our information about the planet's *differential rotation.

Transit instrument A telescope specially mounted, so that it can be used for timing the *transits of stars across the *meridian. It moves only in *declination, and always points to the meridian. When a star approaches the meridian, it may be seen in the telescope field, and crosses a series of wires set up at the focus of the object-glass, so that the moment of transit may be timed very accurately; nowadays, of

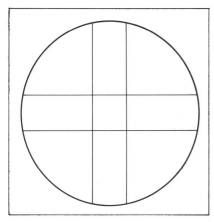

Cross wires of a Transit Instrument.

course, the procedure is carried out electronically. Transit instruments were once all-important for timekeeping. By now they have been to some extent superseded, but transit instruments are still in use. The most modern is at the La Palma observatory in the Canary Islands; it is Swedish, and is known as the Carlsberg Transit instrument. It is purely automatic.

Trapezium Familiar name for the mul-

tiple star Theta Orionis, in the *Orion Nebula (M.42). The name is due to the arrangement of the four chief components, all of which are easy to see with even a modest telescope. Theta Orionis is responsible for making the Nebula shine.

Trifid Nebula Usual name for Messier 20, a gaseous nebula in Sagittarius. It is crossed by dark lanes, and when photographed with a large telescope it is a beautiful sight. M.20 is a typical *emission nebula.

Triple-Alpha Process (*Salpeter process) Nuclear reaction, in which three helium nuclei (alpha-particles) combine to form one carbon nucleus.

Triple point The temperature and pressure at which a substance can exist simultaneously as a solid, liquid or gas—as with H_2O on Earth (ice, liquid water, or water vapour). The conditions on the surface of *Titan may be close to the triple point of methane.

Troilus A *Trojan asteroid, No 1208, discovered in 1931 by Reinmuth. Its period is 11.7 years, and as the mean opposition magnitude is 16 it is not an easy object to observe.

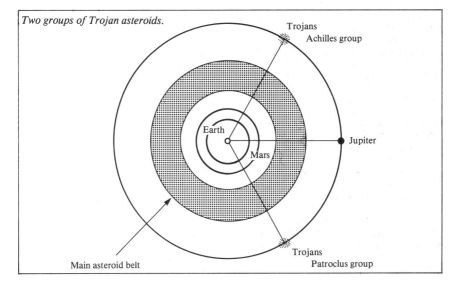

Two groups of Trojan asteroids.

Trojans
Achilles group

Earth

Mars

Jupiter

Trojans
Patroclus group

Main asteroid belt

Trojans Clusters of *minor planets moving in almost the same orbit as Jupiter, the Achilles group leading by about 60 degrees, the Patroclus group trailing by the same amount (see diagram on previous page).

Tropical year This is described under the heading *Year.

Troposphere The lowest part of the Earth's atmosphere, reaching to a height of about 7 miles on average. It includes most of the mass of the atmosphere, and all normal clouds lie within it. Above it, separating it from the *stratosphere, is a shallow layer known as the *tropopause*.

Trümpler, Robert (1886-1956) Swiss astronomer, who spent much of his career in the United States. He concentrated upon studies of star-clusters, and was the first to draw attention to the absorption of light by interstellar matter.

Tunguska Event On 30 June 1908 an object impacted in the Tunguska region of Siberia, blowing pine-trees flat over an

The crater Tycho photographed by one of the Surveyor probes.

area of around 20 square miles. No meteoritic débris was ever found, and no crater was produced. It is now believed that the object was either the nucleus of a small comet, or else a fragment of a comet (possibly *Encke's), which, being mainly ice, would evaporate quickly. It is fortunate that the impact occurred in uninhabited country; if it had hit a city, the death-roll would have been very high.

Tuttle's Comet A comet with a period of 13.9 years; the last return was that of 1980. It was discovered by P. Méchain in 1790 and recovered by H. Tuttle in 1858; since then it has been missed at only one return, that of 1953. At its best it can reach naked-eye visibility, though more generally it is a telescopic object.

Twilight, Astronomical The state of illumination when the Sun is below the horizon, but by less than 18°. In Britain, near midsummer, twilight lasts all night.

Twinkling The popular name for *Scintillation.

Tycho The main ray-crater of the Moon, 54 miles in diameter, with high walls and a central peak. *Surveyor 7 landed on its outslopes in January 1968. Near full moon, the Tycho rays dominate the whole lunar scene, and extend for hundreds of miles in all directions, but they do not issue from the centre of the crater; they are mainly tangential to the walls.

Tycho Brahe (1546-1601) Great Danish astronomer, who was certainly the best observer of pre-telescopic times. Between 1576 and 1596 he worked at his observatory, *Uraniborg on the island of Hven. Using measuring instruments such as *quadrants, made by himself, he compiled a star catalogue which was much better than any previously drawn up, and was subsequently used by his last assistant, Johannes *Kepler, to prove that the Earth is in orbit round the Sun—something which Tycho himself could never accept. Tycho also made observations of the positions of the planets particularly Mars, without which Kepler would never have been able to prove that the Sun is the centre of the Solar System. Tycho was haughty and tactless (his observatory even included a prison, to hold his tenants who refused to pay their rents!) and finally he was compelled to leave Denmark, ending his career in Prague as mathematician to the Holy Roman Emperor, Rudolph II.

Tycho's Star The *supernova of 1572, observed (though not actually discovered) by *Tycho Brahe. See *Supernova.

Tychonic System A theory, supported by *Tycho Brahe, that the planets move round the Sun, but the Sun moves round the Earth. It never attracted much support.

Tyrrhena Terra A dark region on Mars; latitude 10°S, longitude 280°. It lies between Hesperia and Ausonia, and was formerly known as the Mare Tyrrhenum.

U

U Geminorum stars See *SS Cygni stars.

UBV system Colour determinations of stars in the ultra-violet (U), blue (B) and visible (V) regions of the *electromagnetic spectrum. The instrumentation consists of a *photomultiplier together with a set of coloured filters. The intensity of the starlight at various wavelengths can then be used to measure the star's *colour indices in various combinations such as B-V or U-B. This yields important information about the star's surface temperature.

UFOs Unidentified Flying Objects—also called *Flying Saucers. They are due to natural phenomena.

UKIRT (United Kingdom Infra-Red Telescope) It has a 150-in mirror, and is sited on the summit of *Mauna Kea, in Hawaii. It was built for infra-red work, and since infra-red observations do not need optics as accurate as those for visual work the UKIRT was given a lightweight mirror and mounting, but it has proved to be so good that it can also be used in the visual range. At its altitude of about 14,000 ft it is well above most of the atmospheric water-vapour which blocks out infra-red radiations from space.

UK Schmidt The large *Schmidt telescope at the *Siding Spring Observatory in New South Wales. It has a 48-in correcting plate.

UV Ceti stars *Flare stars. UV Ceti itself is the most famous member of the class.

Uhuru (From the Swahili for 'freedom') The first X-ray satellite, launched in December 1970. It provided the first detailed X-ray map of the sky, and discovered X-ray binaries. It finally ceased transmitting in 1973.

Ultra-violet Electromagnetic radiation which has a wavelength shorter than that of violet light, and so cannot be seen with the eye; it lies between the visible and the X-ray range, between 4,000 *Ångströms down to about 100 Ångströms.

Ulugh Beigh (1394-1449) Tartar astronomer, who founded a major observatory at his capital of Samarkand and compiled a good star catalogue. With his assassination, by his son, the Arab school of astronomy came to a virtual end.

Umbra 1. The dark central part of a sunspot. 2. The main cone of shadow cast by the Earth; see *eclipses, lunar.

Umbriel The second satellite of Uranus. It was discovered in 1802 by William *Herschel, but nobody else could confirm it at the time, and it was rediscovered in 1851 by *Lassell. Umbriel is the smallest of the four principal satellites of Uranus; it is believed to consist of about 55 per cent water ice and 45 per cent of rock. For data, see *Satellites.

Undarum, Mare (The Sea of Waves) A small lunar sea, north-east of the Mare *Crisium.

Universal time The same as *Greenwich Mean Time.

Universe Space, together with all the matter and energy contained in it—in fact, everything which exists! Whether it is finite or infinite, we do not know. According to *relativity theory, it may be finite but unbounded, which is a concept almost impossible to visualize.

At present the universe is expanding, inasmuch as all the groups of galaxies are moving away from each other. Whether this expansion will continue indefinitely is not known, and depends upon the overall density of matter in the universe. If this is greater than a certain critical value (about three hydrogen atoms per cubic metre of space) the galaxies will finally draw together; if the density is below this value, the galaxies will continue to separate until all the various groups have lost contact with each other. At present it seems that the density may be insufficient, but there is a great deal of 'missing mass', and the problem is still unsolved.

It is assumed that the universe came into existence between 15,000 and 20,000 million years ago in a *Big Bang, when matter and space were created—though exactly how this came about is quite unknown. Expansion began, and from these mysterious beginnings we can trace the story through to the present day, but we have to admit that we are still completely ignorant about the basic fundamentals. The *steady-state theory, according to which the universe has always existed, has been generally rejected; there is also the *oscillating universe theory, in which Big Bangs occur at intervals of perhaps 80,000 million years, but present evidence seems to be rather against it.

If the increase of velocity with distance is maintained, then a galaxy at around 15,000 million light-years will be moving away at the full speed of light, and we will be unable to see it. If this is correct, then there is a limit to the size of the observable universe, though not necessarily of the universe itself.

University of Hawaii telescope An 88 in reflector on *Mauna Kea.

Uraniborg The observatory on the island of Hven, used by *Tycho Brahe; it also included a second installation, Stjerneborg. After Tycho's departure, in 1596, Uraniborg and Stjerneborg were never used again, and fell into decay. Today nothing is left, though a large statue of Tycho has been erected close to the site.

Uranius Tholus Martian volcano; latitude 26°N, longitude 98°. The base is 40 miles in diameter.

Uranus The seventh planet in order of distance from the Sun. For data, see *Planets.

Uranus was discovered in March 1781 by William *Herschel; it had been previously recorded by several astronomers, including *Flamsteed, but had always been mistaken for a star. It is just visible with the naked eye; Earth-based telescopes show it as nothing more than a pale greenish disk. The tilt of the axis is a remarkable 98°, leading to a strange 'calendar' with each pole having a day equal to 42 Earth-years and a night of equal length. At the present time (1986) the south pole is facing the Sun. The reason for this curious state of affairs is

Comparative sizes of Uranus and the Earth.

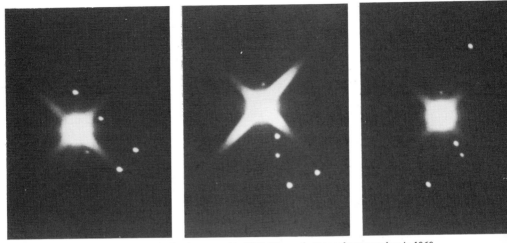

Above *The first five satellites of Uranus, discovered in 1948. These photographs were taken in 1960 and 1961* (McDonall Observatory, University of Texas).

Below *Uranus, showing rings and those satellites known before the Voyager 2 pass.*

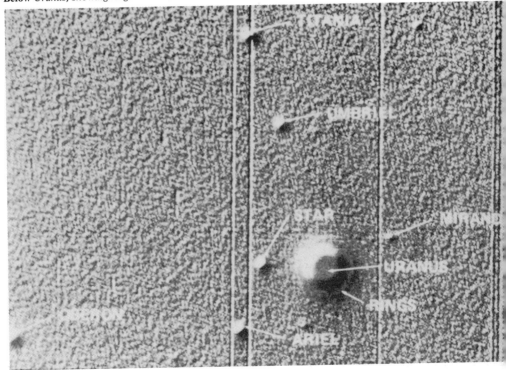

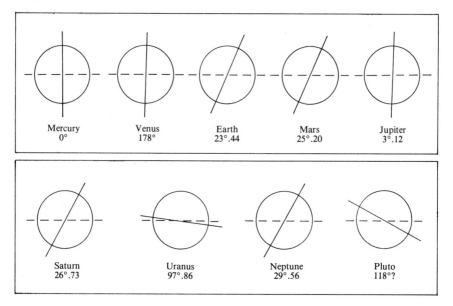

Axial inclination of the planets.

not known; there have been suggestions that in its early history Uranus was struck by an object massive enough to tilt it by more than a right angle. The rotation period of Uranus itself is approximately 16.8 hours.

In 1977 a system of rings was discovered—more or less by chance, during the observation of a stellar occultation by Uranus. The rings are narrow and thin; unlike Saturn's, they are extremely dark, and not easy to record from Earth.

Our knowledge of Uranus was increased beyond all recognition in January 1986, when the Voyager 2 probe flew past the planet at a minimum distance of just over 50,000 miles from the cloud-tops and sent back information from close range. A magnetic field was discovered, together with radiation belts which are stronger than those of Saturn, but the magnetic axis is inclined to the rotational axis by as much as 55°. It seems that there is a substantial core, though with a much

Ring	Distance from Uranus in miles	Eccentricity	Inclination	Width in miles
6	26,030	0.001	0.07	2?
5	26,300	0.002	0.05	2?
4	26,500	0.001	0.02	2?
α	27,800	0.001	0.02	4.3
β	28,400	0.001	0.01	5
η	29,300	0.0	0.0	37
γ	29,600	0.0	0.0	2?
δ	30,040	0.0	0.01	2?
—	31,000	0.0	0.0	?
ε	31,800	0.001	0.0	37

smaller heat-source than for the other giant planets, overlaid by a deep 'ocean' consisting largely of melted water ice together with dissolved ammonia; above comes the atmosphere, largely hydrogen with about 10 per cent of helium. Voyager 2 showed very few clouds, though a few discrete features were made out; there is little detail to be seen on Uranus.

Ten rings are known. Data are as shown in the table.

The ε ring is much the broadest, and is complex in structure, as well as being decidedly eccentric. 'Dust' is distributed all through the ring system, as the Voyager 2 results showed.

Uranus has 15 known satellites. Five (Miranda, Ariel, Umbriel, Titania and Oberon) were known before the Voyager pass; Voyager found another ten, all small, two of which (U7 and U8) are 'ring shepherds'. Ariel has a grooved surface in places, while Umbriel is dark and cratered; Titania and Oberon, also cratered, show signs of past tectonic activity, while Miranda exhibits a varied terrain quite unlike anything else known in the Solar System.

Ursa Major Cluster A *moving cluster. Five of the stars in the *Plough belong to it (all apart from *Alkaid and *Dubhe) and there are many other members, including *Sirius. The apex of the moving cluster lies toward Aquila.

Ursid Meteors A meteor shower reaching its maximum on 22 December; the usual *ZHR is about 12. The parent comet is *Tuttle's Comet.

Utopia A Martian plain north-west of the Elysium volcanic region; latitude 35°N to 50°N, longitude 310° to 195°. It was here that the *Viking 2 lander came down in 1976.

V

V.2 German military rocket, developed at Peenemünde in the late 1930s and early 1940s by a team led by Wernher von Braun. Many V.2s fell on South England during the latter stages of the war, until the launching sites were overrun. Subsequently, captured V.2s were used by the Americans as test vehicles for space research and reached altitudes of up to 244 miles.

VLA (Very Large Array) A major radio telescope installation in New Mexico; there are 27 separate antennæ.

VLBI (Very Long Baseline Interferometry) A method of connecting widely separated *radio telescopes, so as to increase the overall resolving power.

Valhalla The main ringed basin on *Callisto; it is over 370 miles in diameter, surrounded by concentric rings, the largest of which is more than 1,800 miles across. It and *Asgard are much the most prominent features on Callisto.

Van Allen Zones (or Van Allen Belts) Zones around the Earth in which electrically-charged particles are trapped by the Earth's magnetic field. They were discovered in 1958 by James van Allen and his colleagues in the United States, from results obtained from the first successful American artificial satellite, Explorer 1.

There are two main belts. The outer, made up chiefly of *electrons, is strongly affected by events taking place in the Sun; the inner, composed mainly of *protons, is more stable.

Van de Graaff A crater on the Moon's far side, at latitude 27°S, longitude 172°E, notable as being a region of localized magnetic material.

Vaporum, Mare (The Sea of Vapours) A small, very dark lunar sea, near the centre of the Moon's disk as seen from Earth.

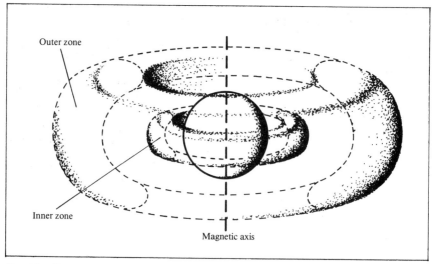

Outer zone

Inner zone

Magnetic axis

Van Allen Zones.

Variable stars Most stars are constant in brightness, remaining unchanged for year after year, century after century. Some, however, brighten and fade over much shorter periods, ranging from an hour or two up to several years. These are the variable stars. There are various types:

1. *Eclipsing variables*. Better termed 'eclipsing binaries', since the stars do not genuinely change, and the apparent fluctuations are due to one component of the binary passing in front of the other. With *Algol stars, there is one main minimum when the bright component is obscured; with *Beta Lyræ stars, the components are less unequal, so that there are alternate deep and shallow minima; with W Ursæ Majoris pairs, the components are dwarfs and almost touching, so that the periods are usually less than a day.

2. *Short-period variables*. Pulsating stars. The *Cepheids have been described under that heading; they are invaluable as 'standard candles'. W Virginis stars, or Type II Cepheids, are also F to G supergiants, but appreciably fainter than the classical Cepheids. *RR Lyræ stars have very short periods, and are almost equal in luminosity; *Beta Canis Majoris and *Delta Scuti stars have very short periods and small magnitude-ranges;

*magnetic variables vary little in light, but have variable spectra. *RV Tauri stars have alternate deep and shallow minima, with periods of complete irregularity.

3. *Long-period or Mira variables*. Red giants, with large ranges in magnitude; neither the periods nor the amplitudes are constant.

4. *Semi-regular variables*. Red giants, with very rough periods, and magnitude ranges much less than for Mira stars.

5. *Irregular variables*. *SS Cygni stars or dwarf novæ, *flare stars, *P Cygni stars, and *recurrent novæ. Some stars, notably *Eta Carinæ, cannot be put conveniently into any class.

6. *Novæ and *supernovæ*. Described under those headings.

7. *Secular variables*. When the alterations are assumed to be permanent—though whether this is so or not is somewhat dubious.

Studies of variable stars are of great importance to astrophysicists. Amateurs can also do valuable work; the method is to compare the variable with nearby comparison stars and work out its brightness, so that after a period a light-curve can be drawn up. Eye-estimates can be accurate to about a tenth of a magnitude, but many modern amateurs use *photoelectric methods, and obtain very precise results.

Variation An inequality in the motion of the Moon, due to the fact that the Sun's pull upon it throughout its orbit is not constant in strength.

The term is also used with respect to the direction of a compass-needle, which will point not to the geographical north pole, but to the magnetic pole. Magnetic variation is the difference, in degrees, between true north and magnetic north.

Vassenius, Birger (1687-1771) Swedish

Light curves of variable stars.

astronomer, who was the first to describe the solar *prominences.

Vastitas Borealis The vast north polar plain extending all round Mars.

Vega The star Alpha Lyræ. For data, see *Stars. Vega is much the bluest of the really brilliant stars; it is just *circumpolar from Britain, and is almost overhead during summer evenings. In 1983 the *IRAS satellite found that Vega is associated with cool material which may be planet-forming or even a planetary

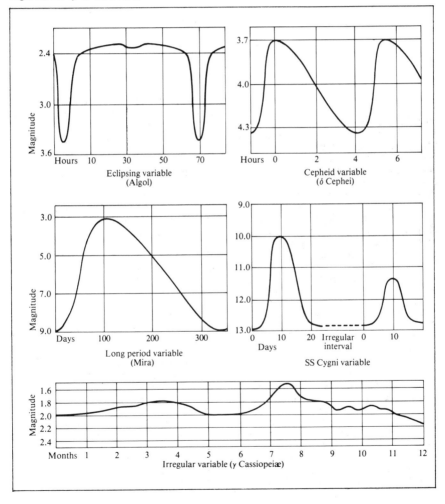

Eclipsing variable
(Algol)

Cepheid variable
(ó Cephei)

Long period variable
(Mira)

SS Cygni variable

Irregular variable (γ Cassiopeiæ)

The first Soviet Venera probe.

system, though the fact that Vega is hotter and more massive than the Sun means that it evolves more quickly—so that even if planets exist (which is by no means certain) it is not likely that life has had time to develop there.

Vela Pulsar The second pulsar to be optically identified. It has a period of 0.089 second, and lies in the *Gum Nebula in Vela. Its optical magnitude is 26, so that it is one of the faintest objects ever recorded. From the rate of its slowing-down, it can be calculated that the supernova outburst may have occurred about 11,000 years ago, and would have been extremely brilliant, though suggestions that it may have been recorded in those ancient times are difficult to take really seriously.

Vendelinus A lunar walled plain, 103 miles in diameter, near the Moon's east limb; it is one of a great chain which includes Furnerius, Langrenus, Petavius and the Mare Crisium. Vendelinus is less perfect than the other members of the chain of similar size, and so is presumably older; its walls are broken and distorted in places.

Venera probes Unmanned Russian space-craft to Venus. Sixteen have been launched, the first in 1961 and the latest two in 1983. Data are as follows:

Vehicle	Launch date	Comments
1	1961	Failure; contact lost.
2	1965	No Venus data received.
3	1965	Landed, but crushed by Venus' atmosphere during descent.
4	1967	Data transmitted during descent to Venus' surface.
5	1969	Data transmitted during descent to Venus' surface.
6	1969	Data transmitted during descent to Venus' surface.
7	1970	Lander transmitted for 23 minutes after arrival.
8	1972	Lander transmitted for 50 minutes after arrival.
9	1975	Transmitted from surface. One picture received.
10	1975	Transmitted from surface. One picture received.
11	1978	Transmitted data for 60 minutes.
12	1978	Transmitted data for 60 minutes.
13	1981	Transmitted data for 60 minutes.
14	1981	Similar programme to Venera 13.
15	1983	Landing pictures; radar mapping by orbital section.
16	1983	Similar programme to Venera 15.

There was some confusion about the earlier results, since it was not appreciated that the data received during the descent of Veneras 4, 5 and 6 referred to the middle atmosphere rather than the true surface. The best results have come from Veneras 15 and 16, which provided high-resolution radar maps of the surface of the planet.

Venus The second planet in order of distance from the Sun. For data, see *Planets.

When seen with the naked eye, Venus is remarkably beautiful, and is much brighter than any other planet or star; at its best it may even cast a shadow. It is at its most spectacular either in the western sky after sunset, or in the eastern sky before dawn; it can never be seen throughout a night, but may at times rise five hours before or set five hours after the Sun. Keen-sighted people can often see it in full daylight, provided that it is known exactly where to look.

Since Venus is almost as large and massive as the Earth, and is closer to us than any other planet, it might be expected to show considerable surface detail. In fact it does not, because the surface is

Phases of Venus, to scale, showing the changing apparent diameter.

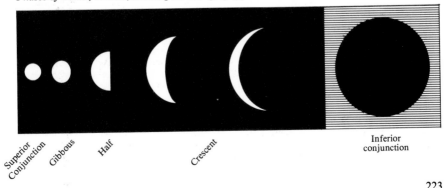

Superior Conjunction   Gibbous   Half   Crescent   Inferior conjunction

The surface of Venus photographed from the descent module of the Soviet Venera-13 probe on 1 March 1982 (Novosti).

permanently hidden by the dense, cloud-laden atmosphere. Generally, nothing is visible telescopically apart from the characteristic *phase, and the vague markings glimpsed from time to time are purely atmospheric in nature.

Before 1962, Venus was regarded as a planet of mystery. Its rotation period was unknown, and nobody could be sure whether the surface were water-covered or completely dry. Spectroscopic analysis had shown that the upper atmosphere is very rich in carbon dioxide, but nothing definite was known about the lower layers. Then, in 1962, the American probe *Mariner 2 by-passed Venus within 21,000 miles, and showed that the surface temperature was much too high for water to exist in liquid form.

Since then there have been many successful missions by both American and Russian vehicles; several *Veneras have sent back surface pictures following controlled landings, while almost the entire planet has been mapped by radar from the orbiting sections of the Veneras

and by the US *Pioneer orbiter. What we have learned changes our view of Venus as a welcoming planet. It is possibly the most hostile world known to us.

The surface includes highlands, lowlands, and a huge rolling plain which extends all round the planet. There are two main highland areas, *Ishtar Terra in the north and *Aphrodite Terra straddling the equator; Aphrodite measures 6,000 × 1,990 miles, and consists of eastern and western ranges separated by a lower region. Ishtar, with a diameter of about 1,800 miles, has the lofty *Maxwell Mountains at its eastern end, rising to almost 7 miles above the mean level of the planet's surface. Of particular interest is the smaller highland of *Beta Regio, which seems to consist mainly of two large shield volcanoes, *Rhea Mons and *Theia Mons. A variable content of sulphur dioxide gas in the atmosphere is an indication that these volcanoes are almost certainly active, and there is another active region, *Atla Regio, adjoining Aphrodite (often nicknamed 'the

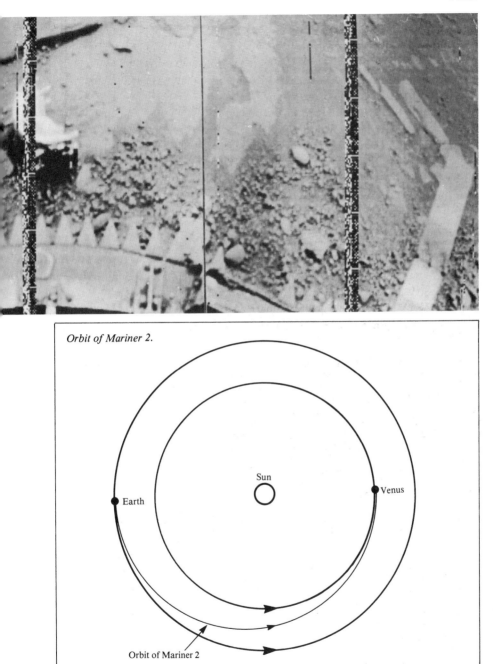

Orbit of Mariner 2.

Sun

Earth

Venus

Orbit of Mariner 2

Above *Mariner photograph of cloud-shrouded Venus.*

Left *The 85-ft radio telescope at the US National Radio Astronomy Observatory at Green Bank, West Virginia* (US Information Service).

Scorpion's Tail', from its shape). Thunder and lightning are probably almost continuous, and it has been said that Venus approximates closely to the conventional picture of hell!

Apparently Venus, unlike the Earth, is a one-plate planet, so that plate tectonics in the terrestrial sense does not apply here, and the volcanoes continue to be active over very long periods, though there are far fewer active regions than on the Earth.

The atmosphere is made up chiefly of carbon dioxide, while the clouds contain large quantities of sulphuric acid. The wind structure of the atmosphere is unique. The upper clouds have a retrograde-sense rotation period of four days, while the planet itself takes 243 days to complete one rotation—longer than Venus' 'year'—and this is also presumably the rotation period of the lower atmosphere. The cloud-deck ends at about 18½ miles above the ground, and the lowest layers of the atmosphere have

227

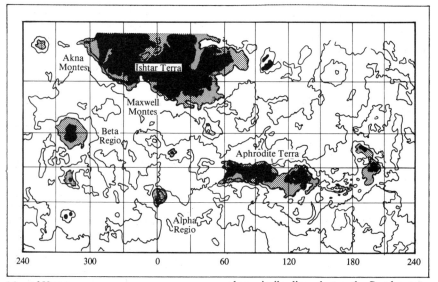

Map of Venus.

been described as superheated, corrosive 'smog'; the surface temperature is over 900° Fahrenheit, while the atmospheric pressure at the surface is over ninety times that of the Earth's air at sea-level.

The axial inclination is 178°, so that the rotation is retrograde. If the Sun could be observed from the planet's surface it would rise in the west and set in the east; the solar day is 117 times as long as ours. But in fact the Sun could never be seen by an observer standing on Venus; it would be permanently concealed by the clouds. The sky would be bright orange, as has been shown by the pictures sent back by the soft-landing Veneras, and though the winds are sluggish they have tremendous force. There is no 'rain'. Evidently the droplets of sulphuric acid condense in the high atmosphere and fall, but evaporate before reaching the surface.

Since Venus and the Earth are so alike in size and mass, we must ask why they are so different in nature. (Venus has no detectable magnetic field, and no satellite.) The answer must lie in Venus' lesser distance from the Sun. It may well be that in the early history of the Solar System, when the Sun was less luminous than it is now, the two planets started to evolve along similar lines; but as the Sun became more powerful, the oceans of Venus boiled away, the carbonates were driven out of the rocks, and the atmosphere became rich in carbon dioxide, which exerts a 'greenhouse effect' and shuts in the Sun's heat. Venus today is certainly unsuited to life of our type, and there is no prospect of sending astronauts there. It has been suggested that in the far future it may be possible to 'seed' the atmosphere, breaking up the carbon dioxide and sulphuric acid and replacing it with free oxygen; in this case the temperature would fall quickly—but whether anything of the sort can be done is highly uncertain. Certainly it is quite out of the question in the foreseeable future.

Vernal equinox The *First Point of Aries.

Vertical circle A great circle on the *celestial sphere which cuts the horizon at right angles and passes through the observer's *zenith.

Vesta The brightest asteroid, though not the largest. For data, see *Minor Planets. At its best it is just visible with the naked eye. It is classed as an eucritic asteroid, with an *albedo of 0.23 (as against only 0.05 for *Ceres).

228

Viking 1. An early series of American unmanned rockets.

2. The probes which made controlled landings on the surface of Mars, while their orbiting sections continued in paths round the planet, undertaking investigations and mapping while also acting as relays for the landers. Viking 1 landed in the plain of *Chryse on 20 July 1976; Viking 2 in *Utopia on 3 September 1976 (the distance between the two landers was 4,612 miles). Both were extremely successful, and during their active lifetimes sent back data of all sorts, together with pictures direct from the Martian surface.

Virgo Cluster A vast cluster of galaxies, about 50 million light-years away. Its overall diameter is about seven million light-years, and it contains over 1,000 galaxies, of which the most massive is *Messier 87. There is evidence that our *Local Group is strongly influenced by the much more massive and populous Virgo Cluster, and the two may make up part of a *supergalaxy.

Vogel, Hermann Carl (1841-1907) German astronomer, who became the first Director of the Potsdam Observatory. He carried out pioneer work on stellar spectroscopy, and was concerned with the discoveries of *spectroscopic binaries.

Volund An active volcano on *Io, originally known as Plume 4.

Voskhod Manned Russian space-craft, succeeding the *Vostoks. From Vostok 2, in March 1965, Alexei Leonov made the first 'space-walk'.

Vostok Manned Russian orbital vehicles; on 12 April 1961 Vostok 1 carried the first space-man, Yuri *Gagarin, into orbit. There were six Vostoks in all; the last

Viking Orbiter and Lander (JPL).

(1963) carried the first woman cosmonaut, Valentina Tereshkova, who completed 48 orbits in nearly 71 hours.

Vulcan During the 19th century it was believed that there could be a planet moving round the Sun at a distance much less than that of Mercury. The French astronomer U. J. J. *Le Verrier, whose calculations had led to the discovery of Neptune, was convinced of its reality, and it was even given a name—Vulcan. However, its existence was never confirmed, and it is now safe to assume that there is no planet within the orbit of Mercury, though comets and small asteroids may approach closer to the Sun.

W

W Ursæ Majoris variables Eclipsing binaries with periods of a few hours. Both the components are dwarfs, and are almost in contact, so that each is distorted tidally into an ellipsoidal shape.

W Virginis variables Type II *Cepheids. The most celebrated member of the class is the southern Kappa Pavonis, which has a magnitude range of from 3.9 to 4.8 and a period of 9.1 days. It is too far south to be visible from Britain.

Walled plain A term applied to the larger lunar walled formations which are often craters.

Walter An 80 mile lunar formation at the edge of the Mare *Nubium. It is one of a chain of three major craters, the others being Regiomontanus and Purbach.

Waqar A large pair of impact craters in Arabia, discovered in 1932. The diameter of the larger crater is over 300 ft.

Wargentin A lunar plateau near the Moon's south-west limb, 55 miles in diameter, unlike any other lunar formation of comparable size; it is in fact a crater filled with lava almost to the brim. It is named after the Swedish astronomer Per Wargentin (1717-1783), a specialist in planetary motions.

West's Comet A bright comet seen in 1976, discovered by the Danish astronomer Richard West. It was a prominent naked-eye object for several mornings, but showed obvious signs of breaking up after perihelion. Its period has been calculated as 558,300 years.

Westerbork Radio Astronomy Observatory A major observatory at Westerbok, near Hoogeveen in Holland. There are twelve 25 m 'dishes', ten fixed and two movable.

Wezea The star Delta Canis Majoris; for data see *Stars. It is exceptionally luminous.

Whirlpool Galaxy M.51 (NGC 5194), a spiral galaxy in Canes Venatici close to *Alkaid in the Great Bear. It is face-on, so that its form is well displayed; it is 37 million light-years away, and as its magnitude is 8.1 it is not a difficult telescopic object, though telescopes of at least 15 in aperture are needed to show the form at all clearly. It was the first spiral to be recognized as such, by Lord *Rosse in 1845.

White Dwarf A very small, very dense star which has exhausted its reserves of nuclear power (see *Stars). White Dwarfs are common in space, but their low luminosity means that they are difficult to detect unless they are comparatively close to us. The most famous example is the companion of *Sirius.

Widmanstätten patterns If an iron *meteorite is cut, polished and then etched with acid, characteristic figures of the iron crystals will appear. These are known as Widmanstätten patterns, and are not found except in meteorites.

Wien's Law The law according to which the wavelength at which radiation from a

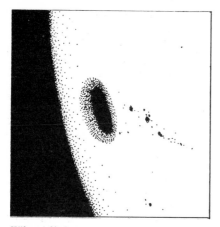

Wilson effect.

luminous source is most intense is inversely proportional to its absolute temperature.

Wild Duck Cluster Nickname for the lovely open cluster Messier 11, in Scutum.

Willamette Meteorite The largest meteorite so far found in the United States. It weighs 14 tons, and is on display at the Hayden Planetarium in New York.

Wilson effect When a sunspot is close to the Sun's limb, it will be foreshortened, so that a spot which is actually circular will apear elliptical. In 1769 the Scottish astronomer A. Wilson pointed out that in general, the *penumbra of the spot will appear narrower in the direction toward the Sun's centre than toward the direction of the limb, and he concluded from this that spots must be saucer-shaped depressions. If the spots were elevations, then the penumbra would appear narrower toward the limb.

Wilson-Harrington Comet A comet discovered in 1949. Its period was calculated as 2.3 years, which would make it the shortest known, but the comet has never been seen again.

Window A region of the *electromagnetic spectrum to which the Earth's atmos-

phere is transparent. The two main windows are the optical (3,000 to 10,000 *Ångströms) and the radio (a few millimetres to about twenty metres).

Winter solstice This is described under the heading *Solstices.

Wolf, Max (1863-1932) German pioneer of astronomical photography. He discovered several comets, more than 200 asteroids, and various other objects, including the *North America Nebula. He was the first to detect *Halley's Comet at the 1910 return. His full name was Maximilian Franz Joseph Wolf, but he was always known as Max Wolf. In 1896 he became Director of the Königstuhl Observatory, near Heidelberg, and retained this post until his death.

Wolf number See *Zürich number.

Wolf-Rayet stars Exceptionally hot, unstable stars whose spectra contain *emission lines. See *Stars.

Wright, Thomas (1711-1786) English philosopher, who suggested that the Milky Way system was flattened in shape, and that there were many external systems, though it is true that many of his other theories were very wide of the mark, and he was concerned largely with the religious aspect. It is however worth noting that he believed Saturn's rings to be made up of particles rather than solid or liquid sheets.

Wrinkle ridges Ridges on the Moon's surface, mainly confined to the maria although they may extend on to the highlands as scarps. They rise in places to several hundred feet. They seem to result from faulting or from fissure eruptions, and are often associated with *rills.

X

X-ray astronomy The X-region of the *electromagnetic spectrum extends between 100 and 0.01 Ångströms, between the ultra-violet and the gamma-ray regions. X-ray astronomy could not begin until 1962, with rocket observations, because X-rays from space are blocked out by the Earth's upper atmosphere, and cannot penetrate to ground level. The first discrete source discovered was Scorpius X-1; the second, Taurus X-1, located in 1963, proved to be none other than the *Crab Nebula.

Many new sources were discovered with the *Uhuru satellite (1970). Since then there have been various other satellites. One was Copernicus, which proved to be extremely successful, and then the *Einstein Observatory, which operated between November 1978 and the spring of 1981, moving at a mean height of 500 miles above ground level.

The Sun is a source of X-rays, particularly from the short-lived, violent *flares, but most sources lie far beyond the Solar System. For instance there is *Cygnus X-1, made up of the supergiant star HD 226868, with a mass thirty times that of the Sun, together with an invisible companion of fifteen solar masses which is probably a *black hole; the material about to pass over the *event horizon of the black hole is intensely heated, making it emit X-rays. We also know X-ray binaries, such as *Hercules X-1, and X-ray bursters, which are temporary, intense sources; at the moment their origin and nature are not definitely known. Some stars are known to be sources; thus in 1978 the satellite HEAO 1 (High Energy Astronomical Observatory 1) detected modulated X-rays from the dwarf nova *SS Cygni.

An X-ray telescope makes use of what is termed grazing reflection, in which the incoming radiations are skimmed off the surface at a low angle, pass down a gold-plated tube and enter special detectors, where they can be analyzed and the results transmitted to the ground.

Y

Year The time taken for the Earth to go once round the sun; in round numbers, 365 days. Astronomically, however, there are several kinds of 'years'.

The *sidereal year* (365.26 days, or 365 days 6 hours 9 minutes 10 seconds) is the true revolution period of the Earth.

The *tropical year* (365.24 days, or 365 days 5 hours 48 minutes 45 seconds) is the time-interval between successive passages of the sun across the *First Point of Aries. The First Point is not quite stationary; due to *precession it shifts slightly, which is why the tropical year is approximately twenty minutes shorter than the sidereal year.

The *anomalistic year* (365.26 days, or 365 days 6 hours 13 minutes 53 seconds) is the interval between successive *perihelion passages. It is slightly longer than the sidereal year because the position of the Earth's perihelion in its orbit moves by about 11 seconds of arc annually.

The *calendar year* (365.24 days, or 365 days 5 hours 49 minutes 12 seconds) is the mean length of the year according to the *Gregorian calendar.

We also have the so-called *cosmic year, which is the time taken for the Sun to complete one journey round the centre of the Galaxy: approximately 225 million sidereal years.

Yerkes Observatory Major American observatory at Williams Bay, Wisconsin. It has various large telescopes, among them the 40 in refractor, which was completed in 1897 and remains the largest refractor in the world.

Young, Charles Augustus (1834-1908) American astronomer who specialized in studies of the Sun, and was the first to describe the *reversing layer.

Z

Z Andromedæ variables Otherwise known as symbiotic variables. Such a system is probably made up of a hot dwarf star together with a red giant. The components are close together, and there are complicated spectral changes which are not at all easy to interpret. Z Andromedæ itself has a magnitude range of from 8.3 to 12.4. Other stars of the same type are R Aquarii, AG Pegasi and V Sagittæ.

Z Camelopardalis variables Variables which are in many ways similar to dwarf novæ or *SS Cygni stars, but differ inasmuch as they experience occasional 'standstills', remaining more or less constant in brightness for a protracted period before resuming their normal behaviour. Other examples are RX Andromedæ, TZ Persei and CN Orionis. Z Camelopardalis itself is the brightest member of the class, reaching magnitude 10.2 at maximum and never falling to as low as 14; its 'normal' period is about twenty days.

Zach, Baron Franz Xavier von (1754-1832) Hungarian amateur astronomer, who made careful studies of planetary movements, and was one of the *Celestial Police. He directed the Seeberg Observatory, in Gotha, and did much to promote international co-operation among astronomers.

Zeeman effect The splitting of spectral lines emitted from a luminous source which is in the presence of a magnetic field.

Zelenchukskaya, Mount Site of the observatory in the Caucasus containing the world's largest reflector, the Russian 236 in (which, to be honest, has never been a real success, and has so far produced surprisingly little work of real value).

Zenith The observer's overhead point (altitude 90°).

Zenithal Hourly Rate When a *meteor shower is being watched, an observer will almost certainly miss some of the meteors, because the *radiant of the shower will not be directly overhead. A correction is therefore applied to obtain the Zenithal Hourly Rate or ZHR, defined as the number of meteors from the shower which would be seen, per hour, under ideal conditions with the radiant directly overhead. Since these conditions are never attained, the observed hourly rate will always be appreciably less than the ZHR.

Zenith distance The angular distance of a celestial body from the *zenith.

Zenith telescope (or Zenith sector) A telescope which is mounted on a vertical axis equipped with a micrometer to measure *zenith distances. To fix the latitude of a position, two stars, one north of the zenith and the other south, are

The Zodiac.

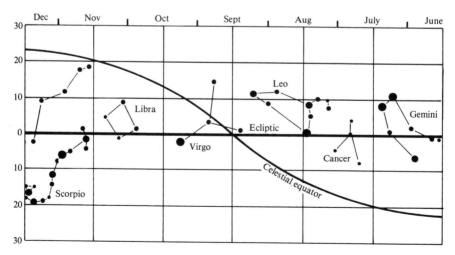

observed as they cross the *meridian. Latitude will then be half the sum of the declinations of the stars, plus half the difference in their zenith distances.

Zero Age Main Sequence (ZAMS) The position in the *Hertzsprung-Russell Diagram where a star first begins to shine by the hydrogen-into-helium reaction.

Zeta Aurigæ An eclipsing binary in Auriga, close to Capella in the sky, making up one of a triangle of three stars nicknamed the Hædi or 'Kids' (the other two are Eta and *Epsilon Aurigæ). Zeta is still sometimes known by its old proper name of Sadatoni. It consists of a hot B-type star and a giant K-type companion. The period is 972 days, and the magnitude range is 3.7 to 4.2. During the partial eclipse of the B-star, its light shines through the outer, rarefied layers of the K-type companion, and the spectral changes are both complicated and informative. Zeta Aurigæ is over 500 light-years away.

Zeta Geminorum One of the brightest of the *Cepheid variables, with a magnitude range of 3.7 to 4.3 and a period of 10.15 days. It has a faint optical companion. The maximum luminosity is well over 5,000 times that of the Sun, and the distance is over 1,400 light-years. Zeta

Geminorum is still occasionally called by its old proper name: Mekbuda.

Zeta Ursæ Majoris See *Mizar.

Zodiac A belt stretching right round the sky, 8° to either side of the *ecliptic, in which the Sun, Moon and all planets apart from Pluto are always to be found. It passes through thirteen constellations, the twelve known commonly as the Zodiacal groups plus a small part of Ophiuchus, the Serpent-bearer. The Zodiacal constellations are:

Aries (the Ram)
Taurus (the Bull
Gemini (the Twins)
Cancer (the Crab)
Leo (the Lion)
Virgo (the Virgin)
Libra (the Scales)
Scorpius (the Scorpion)
Sagittarius (the Archer)
Capricornus (the Sea-Goat)
Aquarius (the Water-bearer)
Pisces (the Fishes)

The Sun, Moon and planets keep near the ecliptic because they move in approximately the same plane; only Mercury has an orbital inclination of more than 5°. *Pluto is the exception, with an orbital inclination of 17°, but, as already noted, it is now very doubtful whether Pluto should be regarded as a proper planet.

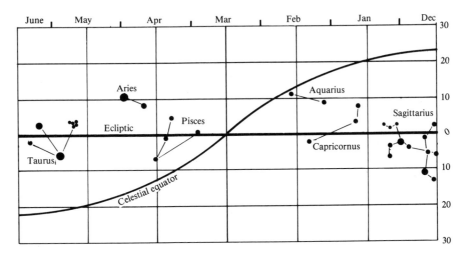

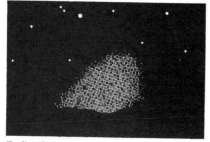

Zodiacal Light.

Aries is always called the first of the Zodiacal constellations; but as the *First Point of Aries has now shifted into Pisces, because of the effects of *precession, Pisces should really have the first place and Aries the second.

Zodiacal Light A cone of light rising from the horizon and stretching along the *ecliptic. It is visible only when the Sun is a little way below the horizon, and from Britain it is never conspicuous, though it may often be observed after sunset in March or before sunrise in September. It is due to small particles scattered near the main plane of the Solar System which reflect and scatter the sunlight. A still fainter extension along the ecliptic is known as the *Zodiacal Band*.

Zöllner, Johann Karl Friedrich (1834-1882) Professor of physical astronomy at Leipzig. He carried out a great deal of useful work, but is probably best remembered for the invention of the *photometer which bears his name.

Zöllner photometer A visual polarizing *photometer. It enables a star to be compared with an artificial star produced in the same field of view, and which can be regulated until its magnitude is the same as that of the star under observation. This is, of course, a measure of the magnitude of the real star.

Zond probes Unmanned Russian spacecraft; eight in the series were launched between 1964 and 1970. Zond 5 (1968) was the first vehicle to circumnavigate the Moon and return to Earth, making a controlled landing; the same procedure was followed with Zonds 6 (1968), 7 (1969) and 8 (1970).

Zone of Avoidance A belt in the sky, centred upon the Milky Way, in which virtually no external galaxies can be seen, because of absorbing material in the plane of our own Galaxy.

Zürich Number (or Wolf Number) A measure of sunspot activity. The Zürich number Z is derived from the formula $Z = k(f + 10g)$, where g is the number of groups, f the total number of individual spots, and k a constant depending upon the experience and equipment of the observer—usually about 1.

Zwicky, Fritz (1898-1974) Bulgarian-born Swiss astronomer who worked in Switzerland before emigrating to America. He specialized in studies of external galaxies, and discovered almost 40 *supernovæ in them.

Index